The Natural History of the San Francisco Peaks

A Sky Island of the American Southwest

The Natural History of the San Francisco Peaks

A Sky Island of the American Southwest

GWENDOLYN L. WARING, PH.D.

The Natural History of the San Francisco Peaks: A Sky Island of the American Southwest
By Gwendolyn L. Waring

First Edition

ISBN 9780692116159
LCCN 2018906304

Published by Gwendolyn L. Waring

Design and production: Mary Williams Design
Front cover photograph: Gwendolyn Waring
Back cover Photographs: Rick Ruess/Fir cones, Ted Grussing/Fall aspens, Max Licher/buttercup, Gary Botello/American Pipit
All figures, unless noted otherwise, by Gwendolyn Waring

Fig. 2.2 Published with the permission of Halka Chronic & Mountaineers Press. From 'Pages of Stone,' 2nd ed., Halka and Lucy Chronic. See reference section.

Fig. 3.1 B. The moraines of the Inner Basin, San Francisco Peaks. Photograph by Troy Péwé, 1968. Courtesy of Lisa and Rick Péwé. From: Péwé & Updike, Museum of Northern Arizona Plateau Vol. 43, No. 2. 1970. See reference section.

Fig. 3.20 Heinlein et al. (2005). Courtesy of Thomas Heinlein et al. and International Journal of Wildland Fire. From Heinlein et al. 2005. See reference section.

This book is dedicated to my mother, Gwendolyn S. Waring,
for her passionate curiosity about things.

Contents

Figures

Tables

Plates

20. *Lupinus argenteus*, silver lupine
21. *Sorbus scopulina*, Arizona mountain ash
22. *Pteridium aquilinum*, brackenfern
23. *Geranium richardsonii*, Richardson's geranium
24. *Pyrola chlorantha*, greenflowered wintergreen
25. *Castilleja miniata*, great red paintbrush
26. *Viola canadensis*, Canadian white violet
27. *Aquilegia chrysantha*, yellow columbine
28. *Erigeron formosissimus*, beautiful fleabane
29. *Maianthemum stellatum*, little false Solomon's seal
30. *Androsace septrionalis*, pygmyflower rockjasmine
31. *Calypso bulbosa*, fairy slipper
32. *Lonicera involucrata*, twinberry honeysuckle
33. *Penstemon whippleanus*, Whipple's penstemon
34. *Carex haydeniana*, cloud sedge
35. *Geum rossii*, Ross's avens
36. *Silene acaulis*, moss campion
37. *Polemoniun viscosum*, sticky polemonium
38. *Packera franciscana*, San Francisco Peaks ragwort
39. *Ranunculus inamoenus* var. *subaffinis*, graceful buttercup
40. *Primula parryi*, Parry's primrose
41. *Gentianella tenella*, annual gentian
42. *Gentianopsis barbellata*, perennial fringed gentian
43. *Luzula spicata*, spiked woodrush
44. *Oxyria digyna*, alpine mountain-sorrel
45. *Boletus edulis*, red king bolete, mycorrhiyza
46. *Calvatia booniana*, giant puffball, saprophyte
47. *Pholiota squarrosa*, Basque garlic, parasite
48. *Usnea hirta*, Old man's beard lichen
49. *Hypnum sp.*
50. *Ceratodon purpureus*

Additional resources that will greatly enhance the use of this book:

- Flagstaff Trails Map, Emmitt Barks Cartography, Flagstaff AZ. This is a great companion for the Peaks book. There are many trails and other features that create a greater understanding of the mountain!

- KachinaPeaks.org: This site has everything you want to know about avalanches on the Peaks, including maps of avalanche paths, avalanche preparedness training, etc.

- SEINET.org: This amazing site includes Will Moir's plant list for the Peaks and surrounding areas, and informs about individual species, including their ranges.

- Ebird.org: a great way to learn about birds on and around the Peaks.

- NAU Pollinator Ecology Group: Visit cpbc.bio.nau.edu

- GrandCanyonWildlands.org/SFPeaksLibrary: The Natural History of the San Francisco Peaks' digital library features many of the documents referred to in this book.

- Springer, J.D., M.L. Daniels & M. Nazaire. 2009. Field guide to forest & mountain plants of Northern Arizona. Northern Arizona Ecological Restoration Institute, NAU, Flagstaff, AZ.

- Kamp, K. 2006. Medicinal flora of the San Francisco Peaks: a historical compilation. Masters Thesis, Northern Arizona University, Flagstaff, AZ.

- https://gapanalysis.usgs.gov/gaplandcover/. Learn about GAP data and mapping here.

- Waring, G. 2011. The Natural History of the Intermountain West. University of Utah Press, Salt Lake City, UT.

- Vankat, J. L. 2003. Vegetation dynamics on the mountains and plateaus of the American Southwest. Springer, NY.

Preface

This is a story about the building of the landscape that encompasses the San Francisco Peaks and its ecological world. This young volcanic mountain is a child of the Pleistocene Ice Age, and its unbelievable history provides a great reminder about the wildness of this world around us, wherever we happen to be.

The San Francisco Peaks is a sky island in the American Southwest. It lies on the southern edge of the Colorado Plateau in northern Arizona. It rises to 12,633 feet as a high and isolated feature on the Plateau, and so many of its tundra and forest plants are separated from related populations by hundreds and hundreds of miles of inhospitable habitat, including the Cold Desert. The Peaks is the only mountain in Arizona that supports plants above treeline, making for a very rare ecosystem here. The many tiny tundra plants above treeline on the Peaks are from the Rockies and even the Arctic. The Peaks is the largest volcano in a field of more than 600 volcanoes.

First a couple of details about this book: The Peaks is a massive volcanic mountain that is crowned by 7 peaks reaching over 11,000 ft (3,350 m). It was once shaped like Mount Fuji, though it was even taller, before it lost its cone. Because the Peaks is a single mountain I refer to it as a singular entity—with a plural spelling—through the book. This is local tradition and personal preference, and, frankly, the alternatives are worse; this did it for me: try saying 'the Peaks is a sky island' versus 'the Peaks are a sky island' or 'the Peaks are sky islands.' Nope, none of it totally works, so I'm going with singular, 'the Peaks is a sky island.' Thanks for your patience. Here's to language and trying to express ideas creatively and honestly.

In order to develop a plant list for the Peaks, I had to draw a line delineating the base of the mountain; I excluded nearby volcanic mountains. An earlier plant list for the Peaks developed by USFS ecologist

William Moir (2005), included surrounding mountains, such as Kendrick Mountain. His list provided the basis for this book's plant list. By focusing on the plants on the Peaks, and excluding those outside of the mountain, a list was born. In the process it quickly became apparent that there are differences between the plant communities on the Peaks and many of its neighboring mountains within the volcanic field, including its next door neighbor Mt. Elden. Elden supports more than 120 plant species that don't occur on the mountain. This wonderful story is still unfolding.

I wrote this story along a timeline, following the Peaks through time as its underlying landmass formed, and then the building of the mountain itself, through a profoundly tumultuous time with lots of tectonic activity and the Pleistocene Ice Age bearing down with endless glacial and interglacial phases—wild weather, plant migrations, unbelievable animals, and then a warming trend that we are in the middle of. I always pull the story back to the Peaks, as data allow, otherwise relying on descriptions of conditions in the closest locations. Of course, this also imparts a greater understanding of the times through the region, especially the Colorado Plateau and the southern Rockies. It's been like working a fantastic puzzle, drawing on research on Kaibab Limestone fossils, radioactive measures of the ages of volcanic rocks, lake cores, caves, packrat middens, glacial debris, layers of sediment that have accumulated in Schultz Creek during the last 6,000 years, and growth rings and fire scars of bristlecone pines and other trees on the mountain. It's been exciting to explore the increasing body of information towards more recent times.

A lot of great research has been done on the Peaks, making for this story. Studies by scientists at Northern Arizona University, the Museum of Northern Arizona, the U.S. Forest Service, the U.S. Geological Survey and the Natural Resources Conservation Service, especially, have helped to provide an understanding of this mountain. My greatest hope is that this book will encourage more research on the Peaks. A greater understanding of this mountain will lead to greater protection. The projects that I long for include understanding what the hell is going on with corkbark fir, and a study comparing the DNA of tundra species on the Peaks with that of related populations in the Rockies and elsewhere, to tell the story of how and when these mountain lovers got here.

I am only just beginning to understand this mountain; this project has been very humbling that way. There is so much more to do. I have tasted the magic of the Peaks. There are dark and mysterious places, kind of primordial forests, which I wasn't expecting in this island surrounded by desert—replete with logs decomposing under a carpet of mosses and lichens, the haunting call of hermit thrushes, and big birds flying fast below dark forest canopies, as on the hike from the Inner Basin to Agassiz Peak.

The position of this mountain island—poised between the vast southern deserts and the mighty Rockies—makes for amazing stories and processes and species, and serious and poignant consideration in the days ahead.

Acknowledgments

Thanks to these people for help, ideas, and conversations about this book

So many people gave so much to make this book happen. It has been humbling..the enormous help and kindness of friends and strangers. I have felt all along that there is a kind of cosmic-ness to this project. Despite occasional setbacks and whatnot, it all just kept coming together in the loveliest way. So, please use this as an inspiration as you aspire to tell stories or whatever…it will work out!

Extra special thanks to Richard Hereford and Max Licher and John Vogel for essential and massive help with the water section, the Peaks' plant list, and the Colorado Plateau and Peaks vegetation maps, respectively. Couldn't have done it without your help. USGS geologist Richard Hereford analyzed available water data for the mountain and was instrumental in organizing that and the climate story. Max Licher made the plant list amazing, with his considerable knowledge of Peaks plants. He also provided most of the plant pictures. Department of Interior geographer John Vogel made the Colorado Plateau vegetation map. He also made a fabulous vegetation map of the Peaks, using 2005 USGS GAP data (https://gapanalysis.usgs.gov/gaplandcover/). His map used 3 dimensional or 3-D data, rather than 2-D, so the areas of the mountain and its different life zones are more accurately represented than ever before! Thank you all.

Huge thanks for amazing conversations, wonderful ideas, guidance, help with data and answers to endless questions about all of it: Gerald Rehfeldt, Tom Kolb, Lee Amoroso, Scott Anderson, Brad Baxter, Sue Beard, Berta Benally, Don Bills, Theresa Clark, Carol Chambers, John Coons, Lee Dexter, Dino DeSimone, Wendell Duffield, David Elliott, Mary Lou Fairweather, Peter Fulé, Monica Gaylord, Heather Green, Gary Hase, Jeremy Haines, Joseph Hazel, Tom Heinlein, Paul Heinrich, Ian Heynssens, Brad Hill, Trevor

Hinckley, Richard Holm, Darrell Kaufmann, Don Keller, Dave Laplander, Thomas Ledig, Mark Manone, Tom McCall, Joel McMillin, Erika Nowak, Michael Ort, Ruthanne Penn, Barbara Phillips, Kirstin Phillips, Peter Pilles, Sue Pratt, Catherine Propper, Nancy Riggs, John Paul Roccaforte, Edward Smith, Susan Smith, John Spence, Larry Stevens, John Spence, Andy Stevensen, Frederick Swanson, Tad Theimer, John Vankat, John Vogel, Bill and Ellen Wade, Jeffrey Walther, Kristen Waring, S.O. Williams, and Erin Young. Thanks to the librarians at the USGS library, NAU's Cline Library, and the Museum of Northern Arizona library.

Very special thanks to Ellen Wade for her tireless editorial efforts.

Thanks to Elizabeth Upham, Ellen and Bill Wade, Sue Pratt, Sue Beard and Richard Hereford, Ian Heynssens, Larry Wood, and Joanie Abbott for hiking with me.

Thank you to so many friends who listened and informed this work: Elizabeth Upham, Ellen and Bill Wade, Sue Pratt, Larry Stevens, Don Keller, Kerry McCracken, Frederica Hall, Star Waring, Dan Johnson, Pat Mehlhop, S.O. Williams, Audria Smith, and Linda Poché.

Thank you, Kelly Burke, Director, Grand Canyon Wildlands Council, for helping to make this happen, and generally being such a great inspiration.

A number of people reviewed the manuscript or portions of it, helping to make it the best it could possibly be: Scott Anderson, Don Bills, Ron Blakey, John Coons, Steven Carothers, Craig Childs, Theresa Clark, Frederica Hall, Richard Hereford, Bradley Hill, Brian Klimowski, Tom Kolb, Chuck LaRue, Michael Ort, Kirstin Phillips, Peter Pilles, Sue Pratt, Catherine Propper, Nancy Riggs, John Spence, Larry Stevens, Ellen and Bill Wade, Kristen Waring, Star Waring, S.O. Williams, and Erin Young.

Financial support came from Geo Fund, Pat Mehlhop, Bob Brill, Thomas Olsen, Ellen and Bill Wade, and many, many others. Thank you for making this possible.

— SECTION 1 —

The Building of the Southwestern United States

The western landscape in which the Peaks exists is profoundly deformed, a sea of mountains and basins, called the Western Cordillera, which extends from Alaska to deep into Mexico (Fig. 1.1). So many aspects of this landscape have made the Peaks what it is today. Here we go.

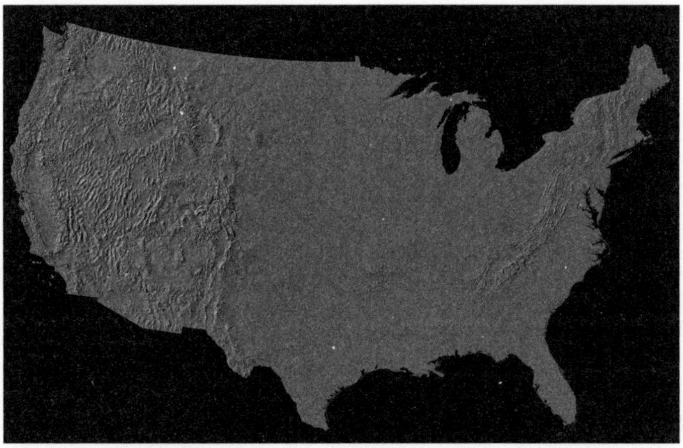

1.1. United States digital elevation map showing how deformed and mountainous the West is relative to the rest of the United States. Can you find the Colorado Plateau, the Colorado River carved into Grand Canyon, and the Peaks? Fig. 3.10 shows the shape of the Plateau. Courtesy of USGS.

This story opens with the last flow of lava all the way around the Peaks, from a tall cone, some 500,000 years ago. This happened during the Pleistocene Epoch, the Earth's last period of ice ages. I like to imagine this volcanic mountain erupting while it was covered with glaciers, the fiery eruption sending molten lava and sizzling ice careening into the sky. It might have gone like that—or it might have gone very differently. The oldest volcano of the Peaks, North Sugarloaf, erupted 2.8 million years ago, slightly before the Pleistocene (Table 1.1). Regardless, this mountain is a child of the Pleistocene. But we need to go back much farther, and consider the land that the Peaks lies on.

Table 1.1. A geologic timeline for the Peaks and surrounding area.
* = marine isotope stage, see figure 2.3, page 30

Period	Epoch	Years ago	Events
	The Present	98	Epic ponderosa pine establishment in the West
		110	Schultz Creek runs after 6,000 years of sediment accumulation; Gustav Pearson arrives to study the Peaks
		115	President Theodore Roosevelt consolidates the Peaks and land around it into the San Francisco Mountains Forest Reserve
		116	John Leiberg evaluates the ecological health of the Peaks and surrounding areas
		128	C. Hart Merriam studies life zones on the Peaks
		130	Fire suppression starts in the Peaks, due to overgrazing
		500	Horses re-introduced into North America by Spanish explorers
		938	Eruption of Sunset Crater, east of the Peaks
		1,000	Sinagua Indians live along Rio de Flag
		2,000	Dry climate replaced by a time of unparalleled climatic variability (Salzer & Kipfmueller 2005)
		6,000	Schultz Creek begins 6,000 years of sediment accumulation, suggesting little precipitation
	Holocene Aridity	8,500-3,000	Dry period, local lakes dry
		9,000-7,000	Oldest record of humans on the Peaks, Bismarck Lake projectile point (Fig. 3.13)
Quaternary	**Holocene**	11,500	End of the Pleistocene
		12,400	Ponderosa pine back around Peaks, higher elevation species migrate back into the Peaks, based on Potato Lake findings
		13,000	Artifacts of Clovis Indians found in northern Arizona; Pleistocene megafauna around the Peaks; horse migrates to Eurasia
	MIS 1 INTERGLACIAL*	11,700 to present	
		26,000	Animas glacier extends 40 miles (~64 km), 100s of feet thick (30 m/100 feet), southern Rockies
	MIS 2 GLACIAL*	27,500-14,040	Closed spruce forest surrounds the Peaks, replacing ponderosa pines, based on Potato Lake fossils
		30,000-13,750	Last glacial maximum: North American glaciers up to 2 miles thick
		35,000-27,500	Mixed conifer forest surrrounds the Peaks, where ponderosa pine forests occur today, based on Potato Lake findings
	MIS 4 GLACIAL*	71,000 - 50,000	
		< 91,000	At least 3 glaciers in the Inner Basin, one leaves glacial outwash on the slopes of Sugarloaf!
		91,000	Sugarloaf Mountain erupts at the base of the Inner Basin
	MIS 5 INTERGLACIAL*	130,000 - 71,000	Plants and animals migrate from Eurasia and down mountain corridors in western North America
		430,000	Eruptions continue on the flanks of the Peaks, including flows from Doyle Peak, Rees Peak, Humphreys Peak and White Horse Hills dome
		500,000	Youngest lavas flow around most of the Peaks' flanks, probably from one tall, central cone

	Pleistocene	2.6 Ma	More than 15 glacial periods between this date and about 12,000 years ago, with the glacial phases accounting for more than 90% of the time
		2.8 Ma	North Sugarloaf Mountain erupts, as the earliest known volcano within the Peaks
		3 Ma	North and South American landbridge allows a plant and animal interchange: tapirs, ground sloths, creosote bush and others make it to North America
	Pliocene	5.9 Ma	
			Sierra Nevada, Cascades and Great Basin form and Colorado Plateau uplifts
		7 Ma	Hopi Buttes volcanoes flowing
		12-13 Ma	Lava flowing on the Mogollon Rim
		18-6 Ma	Playa Basin north of Holbrook/Winslow AZ contains lots of camel and horse fossils from this time
	Miocene	23 Ma	An arid time, expansion of grasslands worldwide, evolution of a North American grazing and browsing mammalian megafauna
		45 Ma	Spruce-fir forests occur in western mountains
	Eocene	56 Ma	Warming climate, subtropical plants make it to Alaska
Tertiary	**Paleocene**	66 Ma	The age of mammals, extinction of dinosaurs
	Cretaceous	145 Ma	Evolution of flowering plants & pollinators; age of dinosaurs
	Jurassic	144 Ma	
		251 Ma	Earliest dinosaurs, with mammals originating shortly thereafter Beginning of the breakup of Pangea into Gondwanaland and Laurasia, and then into modern continents. Evolution of pines before the breakup of Laurasia Kaibab limestone deposited, great diversity of sharks near where the Peaks would be
	Triassic	252 Ma	
		275-250 to 30 Ma	Subduction of the Pacific plate eastward under the western edge of the North American Plate
	Permian	299 Ma	
			Conifers known from the West for the last 300 million years
	Mississippian	359 Ma	
			Devonian rocks laid down below where the Peaks occurs today--during the Mississippian and Permian Periods, lots of fish fossils, including placoderms
	Devonian	419 Ma	The Age of Fishes
	Silurian	444 Ma	First land plants and animals
	Ordovician	485Ma	
			The start of several hundred million years of marine and then terrestrial deposition across the southwestern U.S.
	Cambrian	541 Ma	
		1,700-1,600 Ma	A new section of western North America attaches south of Wyoming
	Precambrian	4,600 Ma	

1.75 billion years ago

Long before the Peaks existed, the western shoreline of North America occurred near northern Colorado. But about 1.75 billion years ago, a new southwestern addition to the continent was forming, as a chain of volcanic islands or arc islands on several tectonic plates collided with the southwestern edge of North America, extending it closer to its current location. These rocks became sutured on to this continent, extending the size of its 'basement' level, made up of metamorphic or cooked rocks including schist, gneiss, and igneous granites. For the next 250 million years, oceans slowly built up the land, by depositing between 2,500 and 5,000 feet (762–1524 m) of marine and coastal deposits of sandstone, shale and limestone over the basement. This was going on right here.

419 to about 360 million years ago

Below the Peaks, below the lava, Devonian rocks were laid down between about 419 and 360 million years ago (Ma) (Table 1.1). Much, much later, along with other marine rocks, pieces of this Devonian layer were pulled back up to the Earth's surface by the viscous lava that formed Mt. Elden, which is right next to the Peaks. It erupted about 600,000 years ago. Big blocks of marine rocks including Devonian, Mississippian, and Permian limestones were pulled and pushed upwards and tilted sideways during Elden's formation (Figure 1.2). You can walk through these marine rock outcroppings on the Heart Trail in nearby Flagstaff. In considering the history of the oceans here, I keep envisioning a vast, beautiful Caribbean-blue ocean, a shallow sea which spanned the Ages of Fishes, and the evolution of amphibians and early reptiles.

Fish were the first animals to have backbones, and they gave rise to amphibians, reptiles, birds and mammals. They are our ancestors! Fishes arose long ago, during the Cambrian, some 500 Ma, and have flourished ever since. Early fishes were jawless, lacking teeth as well as jaws; they lived near the muddy sea floor, sucking up organic debris through their jawless mouths, or filtering small particles from the water. They lacked dorsal and lateral fins, indicating

1.2. Sedimentary rock outcrops at the base of Mt. Elden, pulled up with eruptions of its viscous lava.

that they were not strong swimmers.

And yet jawless fish thrived in Devonian seas. There are many types of them in the fossil record, represented by a variety of shapes and sizes. Some even came to live in fresh water, in rivers and lakes, which set the stage for the movement of animals onto land. At their peak, jawless fish faced overwhelming competition and predation with the evolution of new fish that had fully functional hinged jaws, and teeth. The writing was on the wall and nearly all jawless fish disappeared by the end of the Devonian. In fossil beds in the Baltic region, jawless fish fossils are found with tooth marks in their bony plates; some appear to have healed around the bite marks, suggesting both predation by jawed fish and that they could tolerate it—for awhile, anyway.

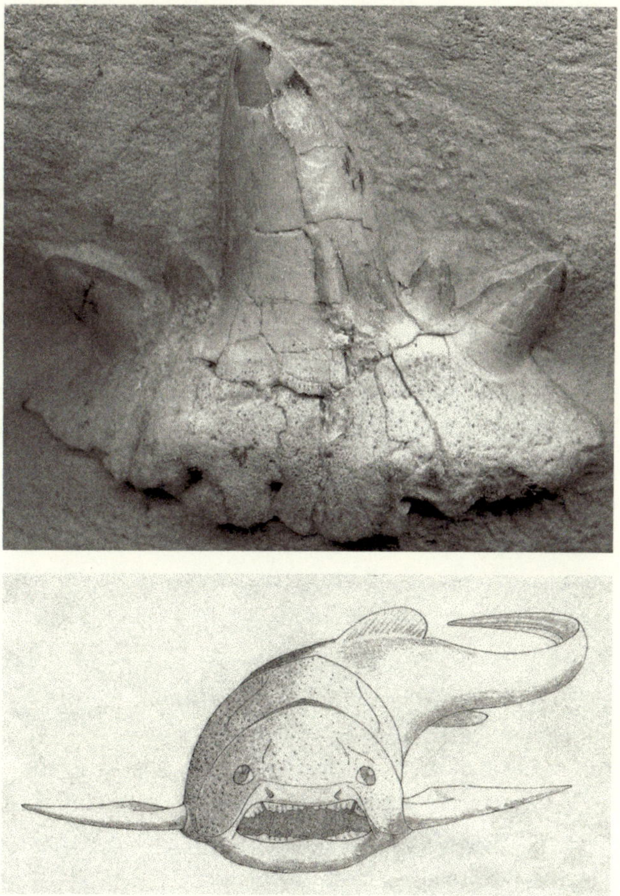

1.3. Top: Fossilized shark tooth from Kaibab Formation near the Peaks. Bottom: A reconstruction of the small placoderm Eldenosteus arizonensis. This placoderm may have been fairly common in the area in the late Devonian. Photo and drawing courtesy of David Elliott, NAU.

The first jawed fish were small, light, spiny, and scaly with large eyes, paired fins and a streamlined shape (Cowen 1976). A later jawed group, the placoderms, was marvelously, heavily armored with a bony shield covering the head and body (Fig. 1.3). Fossils of eight species of jawed placoderms were found in the Devonian rocks near the top of Mt. Elden, representing a great diversity of this group. They were the most savage and powerful fish of the time. Some reached more than 30 feet (9 m) in length, and not surprisingly, dominated late Devonian fish faunas. Predacious sarcopterygians, bony fish with lobe-fins, thought to be the rudiments of limbs, also arose during this time, and ultimately gave rise to four-legged animals—which were the first vertebrate animals to venture onto land! Two sarcopterygians have also been found in the rocks there.

A comparison of Devonian rock outcrops in Arizona near Jerome, on Mt. Elden and in Grand Canyon shows that the area near the Peaks lay at the edge of the Pacific Ocean some 380 Ma (Elliott & Blakey 2005): rocks near Jerome AZ have a great diversity of corals, crinoids, brachiopods, etc., suggesting a shallow marine environment, while Mt. Elden fossils indicate that the water here was even shallower, while Devonian rocks in Grand Canyon are river channels separated by exposed land areas.

Sharks, or cartilaginous fishes, lived alongside the jawed fishes in the Devonian. Their fossils are found in the Permian Kaibab Limestone near the Peaks. This white limestone was laid down between about 280 and 275 Ma, in shallow seas, at the end of the long era of marine deposition that helped build western North America (Elliott and Blakey 2005). Kaibab Limestone can be more than 600 ft (182 m) thick in this region and in Utah. It was the last marine 'transgression,' occurring as the Kaibab Sea washed onto the land, and covered the southwestern portion of the Colorado Plateau with marine carbonate and sand (Blakey & Ranney 2008).

Up to 40 shark species in 8 orders and 18 families have been found in the Kaibab Limestone Formation south of the Peaks, near Flagstaff AZ (Fig. 1.3)! The Kaibab Sea had one of the most diverse shark faunas in the world at that time (Elliott & Blakey 2005, Hodnett et al. 2013)! They lived in salty oceans, and also in freshwater. Their cartilaginous makeup provides only a limited understanding of their history, since cartilage does not persist through time. Their presence in the fossil record is based on their teeth and spines. They have persisted, with some in much the same form, for more than 400 million years, persisting as some of the fiercest predators in today's oceans. The sediment, mostly $CaCO_2$, kept falling out of the ocean, building the southwest landscape even higher, and providing a record for fish in Mississippian and Permian seas. Sharks featured prominently in the ocean through these long periods. Kaibab Limestone fossils include brachiopods, mollusks, bryozoans, corals and sponges.

About 250 Ma

By about 250 Ma, the seas retreated through much of the world . The loss of shallow seas likely contributed to the largest extinction event on Earth, with more than 90% of marine species and 70% of terrestrial vertebrate species being lost. Volcanic eruptions, climate change and/or meteor impacts may have caused this. The building of western North America continued, but now the land was above sea level—it became terrestrial. The rocks laid down at this time were built from blowing sand and sediment deposited by rivers—from continental rather than marine deposition. World-wide processes shaped the new rocks being laid down here. North America began sliding westward away from the supercontinent Pangea. The continental divide of North America lay in the East, along the Appalachian Mountains—they were the tallest mountains on the continent then, having risen when North America and Africa-Eurasia collided a number of times, starting about 480 Ma. Of course, today the continental divide occurs farther west, along the Rocky Mountains. But back then, rivers from the Appalachian continental divide made it all the way to the new dry, southwestern coastal plain that extended from Texas to Utah and Arizona. As eastern rivers reached this flat plain, they slowed down and left mudstone, shale and sand behind, making the Moenkopi Formation, from the early to mid-Triassic, between about 250 and 240 Ma. In places it is more than 400 feet thick (123 m). Ripples formed in these river beds can still be seen today.

The wonderful dark red Moenkopi Formation can be seen jutting out from beneath the lava along Route 66 in Flagstaff. Iron in the sediment oxidized or rusted and gave this formation its reddish color. There are few fossils from this formation. The history of life in these rocks is represented more by tracks than fossils, including those made by reptiles, amphibians and invertebrates. Wupatki National Monument, northeast of the Peaks, has large expanses of Moenkopi at its surface. Native Americans used this rock, which breaks into slabs, for building the many structures there more than a thousand years ago.

After another period of weathering and erosion, other layers of sandstone, shale, mudstone, and volcanic ash were deposited as the Chinle Formation on top of the Moenkopi during a long period of continental dryness and deserts. The Chinle contains the mineral zircon, which ties it to granites in the southern Appalachian Mountains, Texas and Oklahoma, and debris from growing volcanic activity along the west coast (Riggs et al. 1996). The rivers which carried zircon to this area might have been huge, having been compared to today's Amazon River (Blakey & Ranney 2008). This and other layers, here, as at Grand Canyon, have long since eroded away.

65 Ma

Although the building of the larger Western Cordillera has been going on at intervals over the last 300 million years, regional uplift is more recent, occurring during the last 65 Ma (Ronald Blakey, personal communication 2018).

Needless to say, large tectonic processes drove the uplift of western North America. Between 70 and 40 million years ago, the angle of subduction of the Pacific plate moving under the west coast of the North American plate changed: it became shallower. This caused great compression in western North America's landscape. When the Pacific plate previously moved below western NA at a steeper angle (45-60°), it's thought that it was more or less incorporated back into the Earth's mantle (Blakey & Ranney 2008). When the angle of the subducting plate became shallower (~25°), it slid more closely under the land, leading to the Laramide Orogeny, which uplifted the interior of the West, including the Rocky Mountains and the Colorado Plateau. The Colorado Plateau may have been uplifted as much as 15,000 feet or more. While the Rockies and Mogollon Highlands were greatly deformed by pressure from the subducting plate, the Colorado Plateau was only gently deformed with localized 'folding.' By 17 Ma, the Mogollon Highlands collapsed, perhaps due to being stretched to the west by Pacific plates, resulting in its now being at a much lower elevation than the Plateau.

45 Ma

High elevation spruce and fir forests were present in western mountains by 45 Ma, during the mid-Eocene, a time of widespread volcanism (Axelrod 1990). These forests included species very similar to the trees in today's subalpine and mixed conifer forests, including Engelmann spruce, subalpine fir, blue spruce, Douglas-fir and a variety of pines. Of course, conifers are known from the West from much earlier, since the late Paleozoic, with Permian and Triassic (~300 to 250 Ma) rocks of the Colorado Plateau being full of conifer fossils (Ronald Blakey, personal communication).

Spruce probably originated in boreal North America (Ledig et al. 2004). The oldest spruce fossils, found in the Canadian Arctic, date back 45 Ma. They spread through the still-forming North American Western Cordillera during the Miocene Epoch (~23 to 6 Ma), as the new ranges provided cool, montane habitats (Axelrod 1990). It's as though they were ready and waiting to move into the new mountains… to allow them to spread throughout the West. Spruce was found in the Rio Grande Rift by 35 Ma. Later during the Pleistocene, conifers followed the mountains all the way to southern Mexico. Endemic species of Douglas-fir, fir and spruce stayed behind there, diversified and may hold some of the genes that will be needed for whatever climate is coming our way (Ledig et al. 2004).

23 to 6 Ma

The whole world was pretty dry during the Miocene Epoch, between about 23 and 6 Ma. The first grasslands formed as enormous steppes and prairies on nearly all of the world's continents, in areas where forests receded due to aridity. Grasses are more tolerant of dryness than trees are. A great diversity of grazing mammals evolved—many of them quite large—and flourished in the grasslands, demonstrating the profound impact that climate has on the evolution of life (Waring 2011).

As an interesting aside, the evolution of larger body size in mammals had been underway since the end of the Cretaceous, when the large dinosaurs disappeared, resulting in open niches. While Cretaceous mammals weighed just a few pounds, those originating after the Cretaceous-Tertiary mass extinction weighed in at as much as 1,650 pounds (750 kg) initially, and ultimately, Pleistocene species like mammoths and mastodons weighed in at over 22,000 pounds (10,000 kg) (see Pleistocene amimals below)! This large mammal experiment peaked and then ended quickly at the end of the Pleistocene.

Large grazing mammals occurred near the Peaks, too. The diversity of grazing mammals in western and central North America peaked during the mid-Miocene and rivalled the diverse fauna of the Serengeti Plain of Tanzania (Van Devender 1995). As many as 20 species of grazing mammals were found at Miocene grazing sites in the Great Plains. These included camels, rhinos, horses, mammoths, bison, and mastodons. Many also occurred in Arizona, and much of the West. There are abundant fossils of camels and horses in the Hopi Buttes Big Playa, north of the Peaks, dating to between 18 and 7 Ma. Funny thing, in Africa, this loss of forests led to the development of bipedalism in some primates that had previously been tree-climbing quadrapeds.

Humans experienced this megafauna, what a co-existence. Humans made it to the Peaks, and left definite records of their presence nearby, but more about that later (see Pleistocene animals below).

— SECTION 2 —

The Building of the San Francisco Peaks

13 Ma to ~930 years ago

The San Francisco Volcanic Field: The Youngest Geological Feature On The Colorado Plateau

This actually leads us to the volcanoes here and the building of the San Francisco Volcanic Field. The rocks that make up this landscape—up until the lava began to flow here—were mostly made through quiescent forces, the seas passively dispersing sand, coral grains, the remnants of marine organisms, and the like, or by sand blowing in and dunes forming, or rivers flowing through and depositing mudstones and shales and volcanics from all over the continent. And then there were the volcanoes—the youngest geological features on the southern Colorado Plateau.

The San Francisco Peaks resides in a large volcanic field—the San Francisco Volcanic Field (SFVF)—that covers more than 1,800 mi^2 (4700 km^2, Fig. 2.1); it is covered with lava flows from more than 600 volcanoes. While there are volcanoes within the Colorado Plateau, including the Navajo, Abajo, La Sal, and Carrizo mountains, there are even more volcanic rocks along its edge near the Rockies and along its southern edge, the Mogollon Rim (Fig. 2.2; Chronic & Chronic 2004, Betancourt 1990), relating them somehow to the uplifting of the Plateau. Volcanic fields occurring along the edge of the Plateau include the Mount Taylor field in New Mexico, the Springerville field in eastern Arizona, and the Uinkaret field on the north edge of Grand Canyon (Moore 1974).

The SFVF is bounded in the south by the Mogollon Rim, as the southern edge of the Colorado Plateau, to the west by Basin and Range block faulting, to the north by Grand Canyon and the Kaibab uplift, and in the northeast and east by the Black Mesa Basin (Moore 1974). Elevations extend from 12,633 feet (3,850 m) at the top of Humphreys Peak to 4,500 feet (1,370 m) near the Little Colorado River less than 20 miles (32 km) away (Moore 1974).

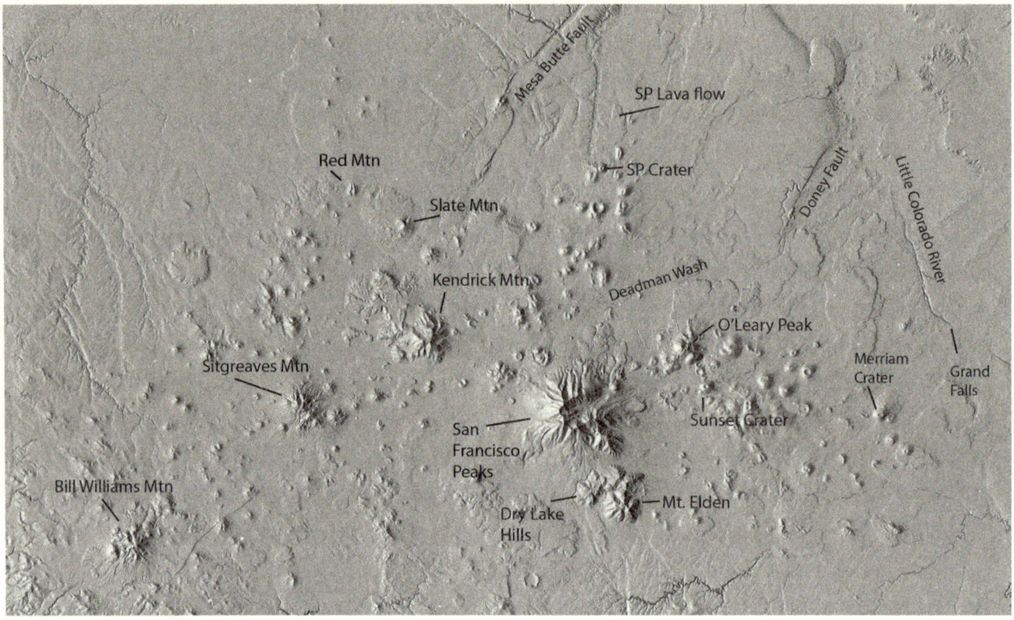

2.1. San Francisco Volcanic Field volcanoes.

Most of the building of the San Francisco Peaks occurred during the Pleistocene, during its many major oscillations of climate between warm and dry interglacial conditions and cold and wet glacial conditions (Fig. 2.3). We don't know how this affected the formation of the mountain, let alone its ecosystems, but I love imagining a collision of ice and molten lava. Pleistocene climates are discussed in greater detail in Section 3 below.

Most volcanic activity occurs at the boundaries of Earth's tectonic plates. The same is true of most of the world's earthquakes. These plate boundaries—where plates meet—are very dynamic places, with one plate pushing up against another, sideways against each other, or sliding under the other plate. Volcanic mountains and lava cover much of western North America and Central America, and yet are often far inland from plate boundaries. The San Francisco Volcanic Field is hundreds of miles from a plate boundary.

It may be that volcanism on the Colorado Plateau represents an after-effect of plate tectonic processes that occurred between billions and tens of millions of years ago (Nancy Riggs & Michael Ort, personal communication 2017). When a volcanic arc island collided with western North America nearly two billion years ago, extending southwestern North America, faulting that resulted from that impact may have persisted in the basement rocks and set the stage for the San Francisco Peaks volcanic field to form… quite a bit later: the collision fractured the land, creating faults through which lava could escape from the Earth's mantle. The San Francisco Volcanic Field overlies several major north- and northeast-trending fault zones. You can actually see cinder cones lined up along some of these faults (Fig. 2.1). Those ancient collisions caused regional faulting that has continued to be expressed through time, enduringly predisposing this region to its volcanic landscape, including

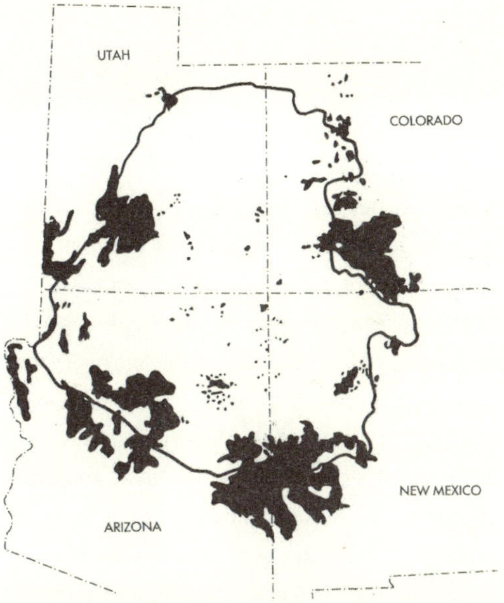

2.2. Volcanic fields along the edges of the Colorado Plateau. Smaller dots are eroded centers of older volcanoes. Courtesy of Halka & Lucy Chronic and The Mountaineers Press.

the San Francisco Peaks, some billion-plus years later—that's crazy.

The formation of the San Francisco Volcanic Field and other southwestern volcanic fields may also be related to the subduction of the Pacific Plate under the North American Plate beginning perhaps 65 Ma. The lower angle, flat-slab subduction glued some of the ocean slab to the North American crust and mantle. When the slab steepened, much later, parts of the hydrated mantle and crust began to 'drip' off. This disruption may have caused volcanism on the Colorado Plateau (Nancy Riggs & Michael Ort, personal communication 2018).

By 12 to 13 million years ago there was lava flowing along the edge of the Mogollon Rim (Fig. 3.10, Plate 3). It has been suggested by Dr. Richard Holm, Northern Arizona University geologist and specialist on the San Francisco Peaks, that since you can walk on lava all the way from Williams south to the older volcanic fields along the Mogollon Rim, they might all be regarded as one volcanic field, which would make the SFVF much larger and nearly twice as old. Other volcanic fields in northern Arizona include the Mormon Lake province (5-10 Ma) extending from Anderson Mesa to Camp Verde, and the Mount Floyd province near Ashfork.

Lava was flowing where the Peaks would develop several million years later. Some of these flows include Switzer Mesa (~5.80 Ma; Updike 1977), Woody Mountain (5.86 Ma), Rio de Flag (6.0 Ma), and Volunteer Canyon (8.68 Ma).

By six million years ago, the development of the San Francisco Volcanic Field in northern Arizona was in full swing, with Bill

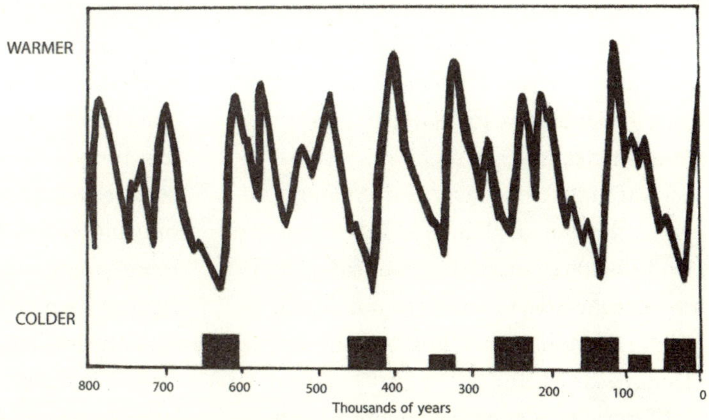

2.3. Marine Isotope Stage, an 800,000-year section of the more than 2 million year climate record established using this method of comparing ratios of stable oxygen isotopes (see Section 3). The blocks at the bottom of the graphic represent glacial periods! See how many there have been since the Peaks came of age. After Railsback et al. (2015). See page 30.

Williams Mountain and volcanoes around it erupting (Fig. 2.1). Generally, the oldest volcanoes in the field, including Bill Williams, occur to the west-southwest, with younger eruptions occurring farther to the east to Sunset Crater, which only erupted about 930 years ago, resulting in an age gradient or a 'younging' of eruptions across the San Francisco Volcanic Field (SFVF). The direction of the younging is actually toward the center of the Colorado Plateau (Michael Ort, personal communication, 2017). In some places the lava flows are quite deep, with the SP Crater flow being 120 feet thick (36 m). Within the Hawaiian Islands volcanic field, the older volcanoes are incrementally older to the northwest, which is the direction in which the Pacific plate is moving. Hawaii's younger, more easterly volcanic islands, Mauna Loa and Kilauea, are still active, and a new island, called Loihi, a submarine volcano, is building towards sea level south of Hawaii, the youngest island.

Volcanoes 101

Before delving farther into the SFVF, here are a few volcano basics. Volcanoes help to cool our warm planet. When Earth formed some 4.55 billion years ago, it may have been a molten ball, made of rocky space debris that was pulled together so forcefully by gravity that it melted into its current form (Duffield 2003). Early on, it was so hot that its core may have resembled a pot of near boiling oatmeal, with a thin, brittle crust (Duffield 2003). Today, the Earth's center is still very hot-its core is thought to be between 9,000 and 11,000 °F (5,000-6,000 °C), keeping much of the core molten. Even several thousand feet down within the Earth's crust, the ambient temperature is still more than 100 °F (38 °C). Though the Earth is cooling, it still remains hot in part because much of its heat is generated by the verrrry gradual decay of radioactive elements such as uranium and thorium.

Volcanoes help Earth to cool by spewing its heat into the atmosphere. As Wendell Duffield, former USGS geologist, says, 'if Earth didn't have a way to get rid of its pent-up heat, it could blow its lid like an overheated sealed container of water' (Duffield 2003). Ouch. So, volcanoes are vents or openings in the Earth's crust that permit heat, magma and gases from the mantle to escape. In the process of dumping heat, volcanoes can be quietly to violently eruptive, depending on the chemistry of the magma, which becomes lava when it reaches the Earth's surface. Thin, easy-flowing lavas are mostly associated with nonviolent eruptions, while volcanoes with thick, or viscous, sluggish-flowing lavas often produce explosive eruptions.

Most volcanic rocks in the San Francisco Volcanic Field are silica-poor and are classified as *basalt* or *andesite*; most cinder cones in the SFVC are made of such lavas. The thickness or thinness of lava is influenced by the amount of silica it contains. This more fluid or less viscous basalt can flow great distances from its eruptive vent to form low, broad volcanic features. Conversely, silica-rich lavas, *dacites* and *rhyolites*, are only locally common, and make up the more mountainous volcanoes, such as the Peaks and the dome volcanoes around it. Highly viscous rhyolite piles up around its eruptive vent to form tall, steep-sided volcanoes, which look nothing like cinder cones. It's easy to imagine the rather blob-like Mount Elden being in the former category (Fig. 2.1).

Water, as steam, contributes greatly to the volatility of volcanic eruptions, as it becomes gaseous and highly pressurized when trapped and heated. Steam in volcanoes, such as Mt. St. Helens, has produced explosive eruptions all over the world. Bubbles of gas or steam escape easily through the thinner lavas associated with cinder cones, and propel a fine spray of magma called a lava fountain into the sky. Dramatic as they may be, sometimes rising to 2,000 ft (610 m), lava fountains are pretty gentle events compared with the eruptions of their silica-rich, rhyolitic counterparts. Rhyolitic magma is so thick that it has a difficult time escaping from the vent, so gas has a harder time escaping, too, further increasing pressure. Such volcanoes often carry blobs of magma, rock and ash along and jet violently upward in hot eruption columns that can rise miles above Earth when they finally erupt. Fine volcanic debris from such eruptions has screened out sunlight on a global scale, sometimes for years (Duffield 2003). Despite such global impacts, volcanoes put out only about a third of 1% of the CO_2 that we modern humans do.

As lava cools and solidifies into rock, some gas remains trapped as bubbles. Almost all volcanic rocks contain some bubbles. Pumice is an especially frothy rhyolitic lava containing lots of silica. The presence of pumice deposits indicates an explosive volcanic eruption. Unlike obsidian, another form of volcanic glass, pumice is foam-like and light-colored; it is extremely porous as steam or CO_2 bubbles became trapped inside as it rapidly cooled and depressurized during eruption.

Cinder cones

The great majority of the 600 volcanoes in the SFVF are cinder cones, and they occur pretty much throughout the field (Fig. 2.1). More than 400 cinder cones were identified and numbered by Harold Colton, Museum of Northern Arizona (Colton 1967).

I love their forms the most (Fig. 2.4, 2.5). They are also the most colorful, often being red or even near purple, indicating tremendous, explosive heat during their eruption. Cinder cones spew cinders and ash. These volcanoes grow when basalt magma erupts in fiery red and orange fountains of lava spray that fall back to Earth to build a cinder cone around the vent. Again, basalt is less viscous or thick because it has a lower silica concentration. Cinders are the basaltic equivalent of pumice.

According to geologist Wendell Duffield (2003), nearly all eruptions in northern Arizona have been relatively non-explosive, most having been of the quiet basaltic-magma variety of cinder cones. And it is cinder cones that have provided nearly all of the great lava flows in our area. Lava flows often ooze quietly from cinder cone vents. One lava flow from a cinder cone near Kendrick Mountain actually flowed all the way to Cameron, a distance of 45 miles (72 km). Lava from SP Crater flowed for some 4 miles (6.5 km), and was 120 ft (37 m) thick in places (Fig. 2.1; Duffield 2003). The lava flow associated with Merriam Crater, another cinder cone, flowed for about 7 miles (11 km), and then poured into the Little Colorado River drainage, damming it up and creating Grand Falls, and then flowed on downstream for another 15 miles (24 km)!

Red Mountain is a cinder cone that may have been partially blasted open through a very explosive event, which formed an amphitheater on one side (Fig. 2.1; Priest et al. 2002). It might have been torn apart by a series of explosive volcanic eruptions caused by intense steam heat. It's thought that cinders falling back on the cone were cemented together and so sealed in gases

2.4. Sunset Crater, 934 years old, the youngest cinder cone in the San Francisco Volcanic Field. Note the extensive colonization of this cone by ponderosa pine.

2.5. Government Prairie cinder cone, west of the Peaks.

trying to escape from the vent. Eventually, the pressure of the superheated steam may have breached the sealed cinders, resulting in one or more steam explosions. Although most cinder-cone volcanoes are piles of loose cinders, the cinders exposed in the walls of the amphitheater at Red Mountain are partly cemented into hard outcrops by a mineralized cement. Also, the cinders range in color from nearly black to reddish and brownish tints, indicating chemical interactions with heated water. The initial amphitheater opening has been enlarged by water and wind erosion since it formed approximately 740,000 years ago.

Lava domes

Lava domes are made of lava rather than cinders or other loose volcanic fragments. As mentioned earlier, their silica-rich lava can be so viscous or thick that sometimes lava domes don't erupt at all, but rather just kind of bulge into place. And yet sometimes they do erupt explosively; Mt. Elden certainly did (see below). One of the oldest volcanoes in the SFVF, Bill Williams Mountain, is a lava dome. It occurs on the western edge of the field and its youngest lavas were determined to be about 2-4 million years old (Table 2.1, Fig. 2.1). Kendrick Mountain, another lava dome, dates to between 1.5 and 3.0 Ma.

Dry Lake Hills and Mount Elden, right at the base of the Peaks, are also lava domes. Their development was largely contemporaneous with that of the Peaks. The youngest lavas in Dry Lake Hills date to about 750,000 years ago, and in Mt. Elden, they are between 500,000 and 600,000 old (Table 2.1, Fig. 2.1). Mt. Elden is a complex lava dome: it has six overlapping lava lobes building out its southern flank, which oozed down from vents high on the mountain. The high viscosity of the lavas forming both of these mountains actually caused massive chunks of sedimentary limestone to be dragged or pushed upward to the flanks of Mt. Elden and along the eastern wall of Dry Lake Hills. In them you can find fossil fishes. Many of the plants growing in the Kaibab limestone wall in Dry Lake Hills along the Little Bear Trail are limestone-loving species, including alumroot and desert columbine. It's funny, this Kaibab limestone wall occurs in a sea of lava, but limestone loving plants found it anyway.

Table 2.1. Approximate ages of some volcanoes in the San Francisco Volcanic Field (Holm 1988, Peters 2003, Morgan et al. 2010, Karátson et al. 2010, Nancy Riggs and Michael Ort, personal communication 2017).

Mountain	Volcano type	Age of lava
Bill Williams	lava dome	2-4 Ma
Sitgreaves	lava dome	2.0-3.0 Ma
Kendrick Mountain	lava dome	1.5-3 Ma
Red Mountain	cinder cone	740,000 (USGS 024-02)
Dry Lake Hills	lava dome	750,000
Mount Elden	lava dome	500,000-600,000
O'Leary Peak	lava dome	270,000
SP Crater	cinder cone	70,000-75,000
Doney Crater	cinder cone	71,000
Strawberry	cinder cone	59,000
Merriam Crater	cinder cone	19,000
Sunset Crater	cinder cone	930
San Francisco Peaks lavas:		
N. Sugarloaf	lava dome	2.8 Ma
Humphreys Pk	composite	
Base/Inner Basin		900,000
Summit		500,000
Agassiz Summit		600,000
Weatherford Canyon summit		300,000
Sugarloaf Mt.	lava dome	91,000

The Mt. Elden eruption started with a powerful explosion that ejected blocks of dense pumice and ash high above its vent and ended with the building of the lava dome, after the gas boiled out. Scattered deposits from this event occur in a debris fan south of Mt. Elden; its pyroclastic flows travelled up to 6 mi (9.65 km) away. Glass and ash remnants of the explosion-the pyroclastic deposit can still be seen at the railroad spur at the Nestle Purina plant—which is 3 mi (4.80 km) to the south.

Another lava dome, Slate Mountain, really brought the difference between lava domes and cinder cones home to me as I hiked it (Fig. 2.1). It is called Slate Mountain because its rhyolite is so gray and platy that early homesteaders believed it to be slate. It is nothing like a cinder cone, though there are cinder cones all around it. Slate Mountain is made up of highly viscous rhyolitic lava, which has lots of silica. Like Mt. Elden and Dry Lake Hills, its thick lavas also pulled a piece of underlying Kaibab limestone to the surface with it, thereby parking older rocks on top of younger rocks.

Sugarloaf Mountain is the youngest lava dome in the field, and the youngest volcano within the Peaks. Its lavas have been dated to about 91,000 years old, and it sits at the lower end of the Inner Basin (Fig. 2.6, Fig. 3.3). Its presence in the Inner Basin tells a great deal about the history and timing of the Peaks' collapse/eruption and its glaciers, but more about that later.

Two lava domes, Bill Williams Mountain and Kendrick Mountain, are thought by some scientists to be stratovolcanoes because they were built from multiple eruptions, so this issue might benefit from further investigation (Robinson 1913, Nancy Riggs, personal communication 2017)!

These volcanoes and Sitgreaves Mountain occur in a line along the Mesa Butte Fault, a major fracture in the Earth's crust that runs nearly 100 miles (161 km) from Chino Valley to Shadow Mountain (Fig. 2.1; Conway et al. 1997), reminding us that all this was made possible by faulting in the Earth's crust.

2.8 Ma to 91,000 years ago

The Building Of The Peaks

The San Francisco Peaks is the star of the SFVF (Fig. 2.1, 2.6). It is built upon a platform of Miocene basalt flows dating around 6 Ma. As a stratovolcano or composite volcano, the mountain was built by many, many volcanic eruptions from nearly 80 vents scattered throughout, over several million years. The oldest volcano within the Peaks is North Sugarloaf, which has been dated to 2.8 million years (Table 2.1). In the process of building the mountain, most flows were buried by progressively younger flows (Fig. 2.7, 2.8, Plate 1; Karátson et al. 2010). The mountain may ultimately have reached an elevation of more than 16,000 ft (4,875 m), compared to its tallest height today of 12,633 ft (3,850 m; Fig. 2.9). The individual peaks on the mountain are the remnants of what was once a Mount Fuji-like stratovolcano, and many of them were also separate vents. The Peaks is the largest stratovolcano or composite volcano and

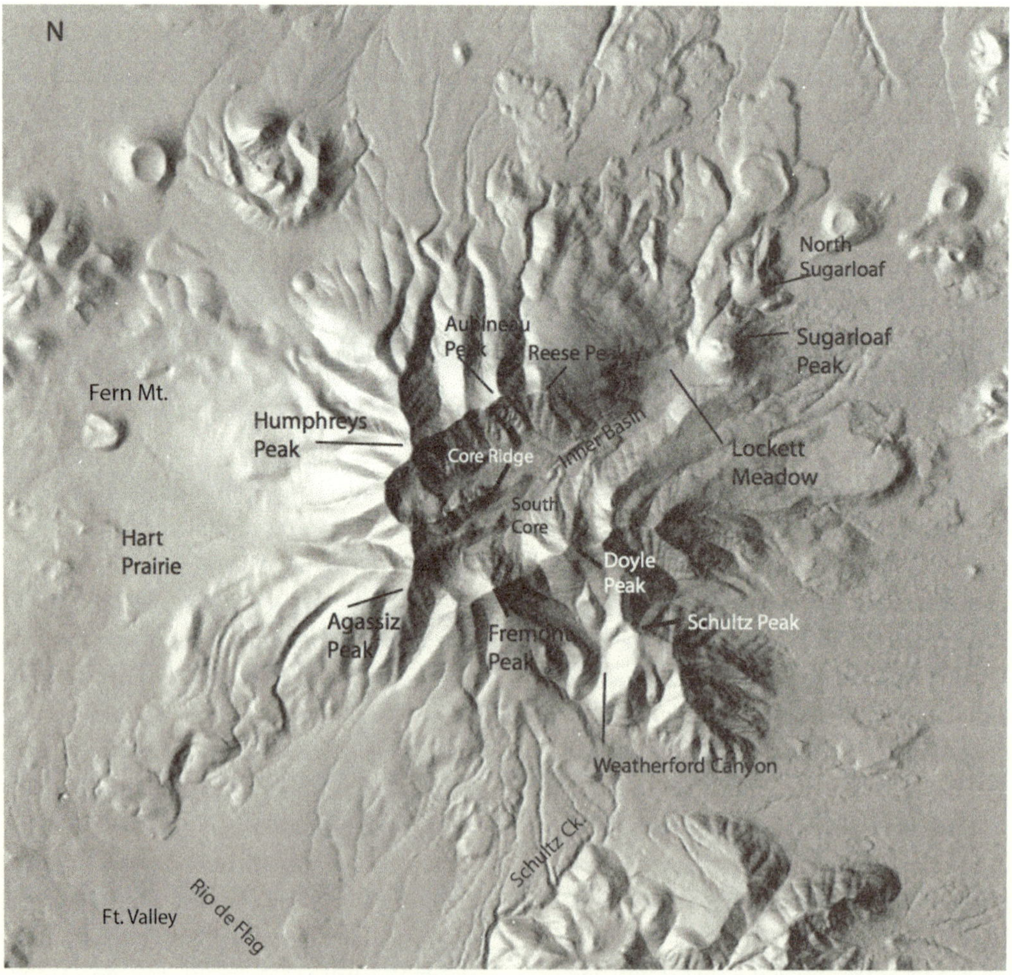

2.6. Peaks closeup, including North Sugarloaf, all Peaks, Sugarloaf, Hart Prairie, Core Ridge, Rio de Flag, Schultz Peak, and Schultz Creek.

the largest volcanic mountain in the SFVF. As beautiful as this mountain is now, it must have been breath-taking in such an iconic, stratovolcano form, before its top collapsed and/or erupted.

There are 7 peaks above 11,000 ft (3,352 m) on the mountain. Six of them are named and include Humphreys Peak 12,633 ft (3,850 m), Agassiz Peak 12,356 ft (3,770 m), Fremont Peak 11,969 ft (3,650 m), Doyle Peak 11,489 ft (3,500 m), Abineau Peak 11,838 ft (3,600 m) and Rees Peak 11,474 ft (3,500 m) (Fig. 2.6). Schultz Peak is 10,083 ft (3,073 m).

This type of volcano consists mostly of alternating layers of lava flows and loose volcanic debris (Fig. 2.7). Such volcanoes are the products of accumulated eruptions of lava and mudflow events that occur over hundreds of thousands to millions of years. The more fragmented layers were formed when loose volcanic debris mixed with snowmelt or rain to produce slurries or mudflows.

Our understanding of the volcanic history of the Peaks is based on extensive geologic descriptions and mapping by geologist Dr. Richard Holm of Northern Arizona University in Flagstaff (1986, 1988, 2004), and many others (Damon et al. 1974, Dennis 1981, Duffield 2003, Karátson et al. 2010, McKee et al. 1998, Robinson 1913, Sheridan and Updike 1975, Updike, 1977, Updike and Péwé 1970, Wenrich-Verbeek 1979). The lavas that make up the Peaks were mapped and aged using potassium-argon dating tests, initially, and later, argon-argon tests and tests for zircons (Table 2.1; Babbitt 1964, Damon et al. 1974, Holm 1988, Morgan et al. 2010, Karátson et al. 2010). The Museum of Northern Arizona, Flagstaff, was instrumental in early Peaks research, with Director Harold Colton providing one of the earliest and most durable classifications of the evolution of the mountain (Colton 1967), and former MNA Director Edwin McKee providing numerous dates for the lavas themselves, which have been invaluable in telling its story (McKee et al. 1998).

The San Francisco Peaks is made up of layers of rather thin andesite, medium-silica lavas, high-silica lava domes with thick blocky flows, and fragmented deposits (Holm 2004). The andesite flows came mainly from central vents, while the thicker lavas erupted on the mountain's flanks and at the base of the central stratocones. Large plugs and dikes in the center of the mountain, including Core Ridge, supplied the stratocones with lava, while lateral dikes also fed flank eruptions (Fig. 2.6).

This composite volcano is made up of at least three tall stratocones, and nine related lava domes. According to Holm (2004), the central part of the Peaks has an eruption history that spans about 1.5 million years, while satellite or flank domes extend its age to

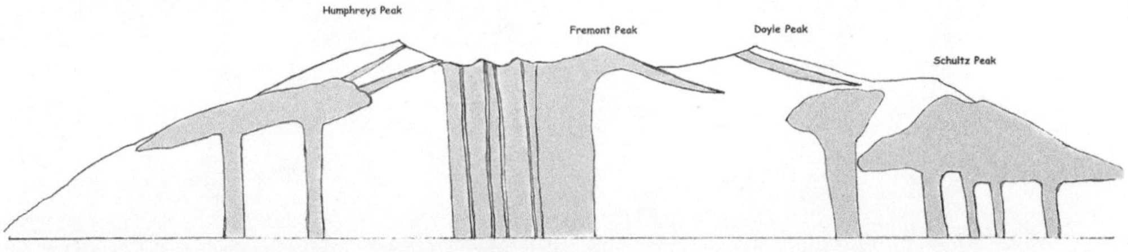

2.7. Some of the lava intrusions within the San Francisco Peaks stratovolcano. After Holm (1988).

about 2.5 Ma. Holm (2004) mentions 79 dikes associated with the mountain; that is, 79 conduits through which magma moved outward to the flanks of the volcano. Most volcanic activity within the Peaks is aligned along north-northwest and northeast-oriented fault zones that underlie the mountain.

According to Holm (2004) and Karátson et al. (2010), the average amount of volcanic material erupted was 7,000 ft^3 (0.2 km^3)

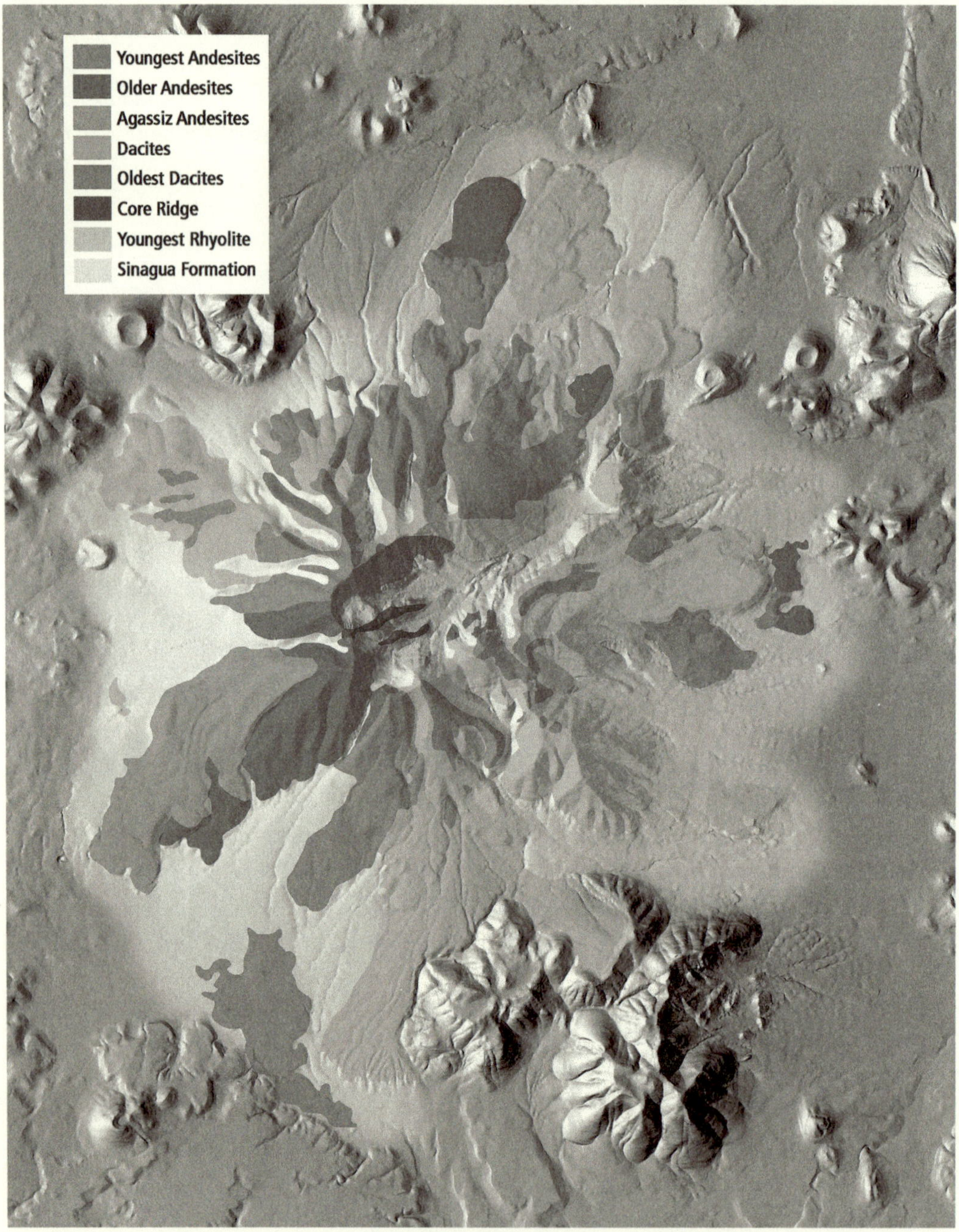

2.8. Distribution and age of Peaks' lavas, including the Sinagua Formation. Note that the youngest andesites (near-darkest hue) occur all the way around the Peaks! After Holm (1988). See Plate 1.

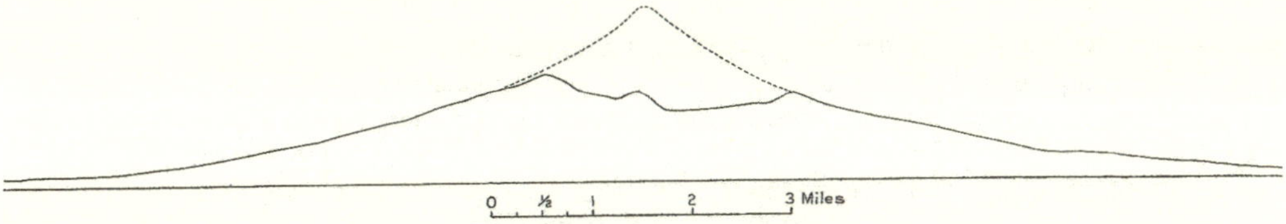

2.9. The Peaks' possible last and tallest cone which released the youngest andesite lava flows all the way around the mountain nearly 500,000 years ago. The Peaks may have been as much as 4,000 feet taller at this time. After Robinson (1913).

every 1,000 years on the Peaks. Karátson et al. (2010) suggest that the final volcanic activity may have increased to a level of 10,000 ft^3 (0.3 km^3) per 1,000 years. This is regarded as a medium development rate for long-lived stratovolcanoes (Karátson et al. 2010). Most eruptions were effusive or flowing, though lava was delivered in more explosive events, as evidenced by the presence of pumice and other hydrothermally-altered rocks.

The Peaks' development is told through the many layers of lava that have been mapped and aged, though many older layers surely are buried under younger lavas. There are also 'rootless' lavas, and even a deposit of pumice on the slope of Agassiz Peak that occurs nowhere else and whose source is unknown. Chemically, most of the rocks making up the Peaks are andesite (85%), which is more viscous than basalt when molten; some are dacite and a few are basalt and rhyolite (Holm 1987). This chemical mix and the interlayering of lava flows and fragmented deposits are all characteristic of stratovolcanoes.

Four phases in the formation of the Peaks

After mapping the lavas of the San Francisco Peaks, Holm (1986, 2004) described three eruption phases in building of the Peaks that culminated in the construction of an andesite stratocone. This series of phases has been proposed by numerous geologists since the early 1900s, including Harold Colton (Colton 1967) and former U.S Secretary of the Interior, Bruce Babbitt (1964).

Each of the first three phases shows the same pattern of dacite and rhyolite lava domes forming, and then being overlain by more fluid andesite lava flows and pyroclastic deposits (Table 1.1). The final fourth phase involved the loss of the mountain's cone and the creation of the Inner Basin.

During the earliest phase of the Peaks' development, some of the oldest lavas were produced by a lava dome, North Sugarloaf, on the northeast edge of the Peaks, some 2.8 Ma. When you drive up the Lockett Meadow Road to the Inner Basin, you are driving along the flanks of this oldest part of the Peaks. A viscous central lava dome, and the Rees Peak and Hochderffer Hills lava domes were also forming and contributing to the building of the mountain early on. During this early phase, between about 2.8 Ma and 900,000 years ago, the mountain is thought to have reached 5,525 ft (1,700 m), *above* the area's baseline elevation of about 7,100 ft (2,164 m)—equaling nearly 12,625 ft (3,848 m), much like its height today (Holm 2004)!

According to Randall Updike (1977), the lower half of Core Ridge and all of south Core Ridge erupted during this early phase (Fig. 2.6). These remain the big volcanic vent features in the interior Inner Basin today. Erupting andesite lavas formed much of the southern part of the Peaks during this time. Andesite lavas also flowed to the southwest, forming the northern boundary of the Rio de Flag near the Museum of Northern Arizona, more than 6 miles (10 km) from the crater. Overall, flows extended farther with this phase than those from later eruptions (Updike 1977).

During the second phase, between about 760,000 and 580,000 years ago, the mountain increased to a height of 6,825 ft (2,100 m) above the baseline, for a total elevation of nearly 14,000 ft (4,270 m)! Flows from a central stratocone and from Fremont peak dome created the south and southeast flanks. Lava flowed from Core Ridge at this time, the Humphreys Peak dome formed on the northwest side, and Schultz Peak dome formed on the southeast side, as did lava domes adjacent to the Peaks, including Dry Lake Hills and Mt. Elden. The whole area was alive with volcanic activity.

During the third phase, between 750,000 and 430,000 years ago, Rees and Fremont lava domes were built. The Peaks may have reached its highest level, growing to nearly 8,125 ft (2,500 m) above baseline, to an elevation of about 15,225 ft (4,685 m). Andesite lava flowed from central stratocones and on the southwest flank of Agassiz. Duffield suggests that the mountain was even taller, possibly reaching an elevation of 16,000 ft (4,900 m; Duffield 2003)!

A Plinian eruption or explosive eruption occurred during the 3rd eruption stage, producing widespread air fall of pumice, and copious ash, and the collapse of part of the volcano. Plinian eruptions are characterized by plumes of volcanic debris and hot gases that may be propelled as high as the stratosphere. The fantastic eruption of Mount St. Helens was also a Plinian eruption. There may have been a lateral collapse of the upper part of the Peaks, resulting in the creation of an ancient valley, perhaps like that of the Inner Basin today.

The Peaks may have reached its greatest height between 514,000 and 505,000 years ago (Karátson et al. 2010). This is suggested by the fact that the youngest andesite lavas nearly encircle the mountain at this time, meaning that they erupted from a central cone high enough to flow down nearly all of its flanks (Fig. 2.8; 2.9, Plate 1; Table 2.1). It is assumed that this was the last eruption of the stratovolcano's central cone. As this lava is absent only from the slopes of Agassiz Peak, it is thought that Agassiz was a tall, resistant, remnant stratocone that blocked the flow of the youngest andesites to the southwest. The last several andesite flows arose through vents in the center of the Inner Basin, just east of Core Ridge (Karátson et al. 2010).

500,000 to 400,000 years ago
MIS 11-13 = Super warm and super cold

So, the eruption/collapse of the Peaks occurred somewhere during super warm and/or super cold phases of the Pleistocene (Fig. 2.3).

During the fourth phase of the Peaks' building, after 505,000 years ago, nearly 4,000 ft (1,200 m) of the upper mountain were lost. It is this part of the story of this mountain, the cause of this massive loss, that is mysterious and delightfully controversial. What was left behind in lieu of the top was a massive basin—the Inner Basin—a long valley that extends through the center of the Peaks and today lies 3,000 ft (914 m) below the mountain's highest summits. It's interesting to consider that there were four glacial episodes recorded in the world, during the last 500,000 years of the Pleistocene (Fig. 2.3, Table 1.1). So, very cold temperatures and glaciers were bearing down on the mountain.

The eruption of the Peaks' youngest lavas of 505,000 years ago and of Sugarloaf Mountain, 91,000 years ago, provide bookends for the timeframe within which the Peaks' cone came down and the Inner Basin formed (Table 1.1, 2.1). This is so because Sugarloaf lava dome lies in the mouth of the Inner Basin where it drains to the northeast. It wouldn't have survived the collapse/eruption of the mountain's top, lying directly in the path of the associated avalanche (Fig. 2.6).

How did the Peaks lose its top?

Many scientists have considered the loss of the Peaks' cone and how it might have gone. Robinson (1913) considered erosion to be the principal process in the formation of the Inner Basin. Today, most geologists conclude that some kind of collapse of the cone occurred (Holm 2004).

These days, the more contentious question is whether the Peaks lost its cone through a violent eruption, much like that of Mount St. Helens, or whether it collapsed into an avalanche—without an eruption at all. Richard Holm (2004) suggests that while most historical lateral-collapse events within volcanoes have occurred in association with magmatic eruptions or hydrothermal explosions, the San Francisco Peaks stratovolcano may instead have failed or collapsed into a 'cold debris avalanche' on the mountain's northeast side, possibly generated by major slope failure, and, importantly, independent of eruptive activity. In this scenario, it may be that associated earthquakes, as at Mount St. Helens, destabilized the mountain. No debris avalanche deposits have been found (Karátson et al. 2010), though they may have been eroded and dispersed downstream, or buried by younger deposits.

Richard Holm (2013) proposes that the Peaks lost its cone in a cold debris avalanche because there is no evidence of an explosive eruption, or any eruption at all. Holm makes the case that stratovolcanoes can be inherently unstable, and for a number of reasons: high, steep slopes increase the likelihood of collapse; top-heavy summit domes above steep slopes can become overloaded and destabilized; cones inflated by magma can oversteepen slopes; younger intruding lavas can fracture and inflate flanks, which may have happened on the northeast side of the Peaks.

According to Holm, no studies of the Peaks have presented evidence for: (1) a hot water system that could have exploded, (2) a lateral explosion deposit, (3) lava extruded into the Inner Basin after its origin, or (4) pumice deposited at the time when the Inner Basin could have originated. According to Drs. Nancy Riggs, NAU geologist, and Wendell Duffield, USGS geologist, a lack of evidence, today, that the Peaks erupted, is not evidence that it did not erupt. Despite this absence, Duffield and Riggs suggest that the Peaks may well have lost its top during a highly explosive eruption, like Mount St. Helens.

The Peaks is not alone among volcanoes in providing few clues about the loss of its cone. The basis for the collapse of a much larger and younger volcano, ancestral Mount Shasta, in California, is impossible to discern. This mountain experienced one of the largest landslides ever recorded, carrying an estimated 10 mi^3 (45 km^3) of debris, much larger than that of either MSH (0.6 mi^3, 2.5 km^3) or the Peaks (2 mi^3, 8 km^3) (Crandell 1989)! Its avalanche is thought to have occurred between 300,000 and 380,000 years ago, making it younger than that of the Peaks. The cause of the landslide is unknown, probably because later flows buried clues. Richard Crandell does not rule out the possibility of the avalanche being triggered by steam explosions within the volcano (Crandell 1989). Earthquakes or even glacial erosion may have triggered the avalanche, though evidence has not been found. Several middle-aged Hawaiian volcanoes, including Haleakala on Maui, also have large central cavities which have generated debate as to whether they arose through erosion or eruptions.

Lessons From Mount St. Helens

Did the Peaks erupt like Mount St. Helens (MSH)? Certainly the two stratovolcanoes are remarkably similar in appearance, with inner basins that open to one side, indicating a lateral failure in the mountain (Fig. 2.10). In addition, a lava dome formed in the MSH inner basin at the end of its major eruption, much like the development of the lava dome Sugarloaf in the base of the Peaks' Inner Basin. While volcanoes were generally thought to erupt through vents from their tops, the eruption of MSH was different: it experienced a massive collapse and explosion on its side: a lateral eruption. The nature of its eruption, and that it was so well-documented, has provided insight into how other volcanoes might have erupted.

For several months from late March in 1980, up to 30-40 earthquakes per day occurred beneath Mount St. Helens, shaking the mountain and surrounding area repeatedly (Swanson et al 2005). During this time, a series of steam explosions blasted two large holes through ice and rock at the mountain's summit and a blister-like bulge grew on the north flank as magma from below filled the volcano. This continued until May 18, 1980, when most of the north flank, precariously over-steepened by the bulging, collapsed. The upper 1,300 ft of the stratovolcano disappeared, reducing its elevation from 9,677 to 8,365 ft (2,950 – 2,550 m), and a north-facing crater or basin formed. The collapse itself created the new basin, removing 0.6 mi^3 (2.5 km^3) of rock from the cone and the basin (Swanson & Major 2005). As the peak collapsed, it exposed a vent housing pent-up gas trapped within the volcano, and so, caused a tremendous explosion to the north. This explosion flattened nearly 250 mi^2 (650 km^2) of mature Douglas-fir forest. Several years after the eruption, a lava dome formed in the crater.

According to Duffield (2003), had such a blast occurred within the Peaks, it would have been even more violent and powerful, as the Peaks and its basin are much larger. Holm estimates that 2 mi^3 (8 km^3) of rock came down from the Peaks' collapse, from the cone and the Inner Basin (Holm 2004). A much larger mountain, indeed.

Nancy Riggs and Wendell Duffield both feel that the physical features of the Peaks could have resulted from a Mount St. Helens'-style lateral eruption and collapse, although they consider it likely that subsequent glaciation also modified the mountain's shape. To uncover the Peaks' story, Riggs is considering dating pumice deposits that occur northeast of the mountain, to see if they are related to the Peaks' loss of its cone.

As Duffield (2003) says, the list of maybes is long, and the answer is not clear. The pyroclastic evidence required to demonstrate an explosion on the Peaks could lie buried under younger lavas in the debris fan to the

2.10. Mount Saint Helens before and after its 1980 eruption. Top: Before. Below: After. Courtesy of USGS.

east or could have eroded away. During the Pleistocene, mountain glaciers and streams modified the shape of the remaining peaks and the Inner Basin (see below), and could have removed evidence of an explosion. Riggs, like Duffield, concludes that we may never know how the Peaks lost its summit. It's fun, though, to imagine an explosion in excess of MSH taking this landscape for a major ride.

Other details of the Peaks' eruptions

Eruptions continued on the northeastern and eastern flanks of the Peaks around 430,000 years ago, including flows from the eastern flank of Doyle Peak and from Rees Peak, a Humphreys Peak cryptodome and dike, and the White Horse Hills dome. The youngest eruption before Sugarloaf occurred around 300,000 years ago as a dome at the top of Weatherford Canyon flowed (Nancy Riggs, personal communication, 2017). This dome is truncated along the rim of the Inner Basin, perhaps by the formation of the Inner Basin or by glaciation. This is yet to be determined. It was still a pretty lively time, even after the top of the Peaks came down!

Sugarloaf Mountain erupts

The eruption of Sugarloaf about 91,000 years ago represents the last known volcanic activity within the Peaks (Fig. 2.6; Peters 2003). According to Updike (1977), its initial eruption was very explosive, as magma encountered groundwater, perhaps from melting glaciers. Being a lava dome like Mt. Elden and Dry Lake Hills, Sugarloaf kind of bulged into shape and was accompanied by a major pyroclastic explosion that left very thick and extensive pumice deposits on the east side of the Peaks. The fallout from this explosive event covered nearly 4 mi^2 (10 km^2), with deposits of ash, pumice and coarser debris reaching a thickness of 30 ft (10 m) or more. These deposits were mined until recently, which is why they are particularly conspicuous.

There is also evidence of glacial outwash deposited around Sugarloaf, as sediment-laden meltwater flowed down the Inner Basin and ground up against its slopes. This happened several times during the last 91,000 years, as glaciers started melting and flowing.

Heat

There is still heat—perhaps considerable heat—associated with Sugarloaf Mountain. Studies of the chemistry of water and travertine deposited with springs on the south side of the Grand Canyon due north of the SFVF contain water from the mantle-derived magma of the San Francisco Volcanic Field (Morgan et al. 2010). This water is from the regional aquifer and suggests that the groundwater is carrying off hot water from the SFVF. This is the reason the Peaks and surroundings have no hot springs. The Sugarloaf eruption, based on its rhyolitic composition and size, should still be quite hot! A USGS study of the geothermal potential between Sugarloaf and O'Leary Peaks suggested that there is enough energy there to power Flagstaff! Such a project would require drilling down into the basement rocks, the Proterozoic rock, below Grand Canyon, to a depth of perhaps 2 miles (3.2 km), to tap this energy source. Well, well, well.

Where to see the different lavas

You can drive through the youngest andesite lavas that flowed all around the mountain along Fort Valley Road near the Schultz Creek turnoff and on up north until the Snowbowl Road turnoff. By contrast, the oldest lavas are encountered when you drive along the outer edge of North Sugarloaf Mountain on the way to the Inner Basin, on the Lockett Meadow Road (FS 552). And, of course, as you drive on up farther, you drive along the flanks of Sugarloaf Mountain—the youngest volcano on the Peaks (Fig. 2.6). Schultz Peak is one of the oldest volcanoes of the Peaks (750,000 years old; Fig. 2.6), and you travel across its flanks when you walk or bike the Waterline Road from Schultz Pass towards the Inner Basin.

Recolonization of the Peaks… Some lessons from MSH

Certainly the type of event that brought down the Peaks' cone had a huge influence on its biological communities. It seems to me that a massive cold debris avalanche (Holm 2004) would have destroyed everything in its path out the northeast side of the mountain, but populations outside of this area of disturbance might have been less affected. However, if the collapse of the Peaks involved an explosive eruption, things might have gone very, very differently. The explosive eruption of Mount St. Helens provides a great example of how such events affect the ecology of volcanic mountains.

Within about 12 hours on 18 May, 1980, the eruption of MSH transformed 230 mi^2 (600 km^2) of lush green forests, meadows and clear cold lakes into a stark gray ash and pumice covered landscape (Crisafulli et al. 2005).

The massive eruption destroyed much of the vegetation on the volcano's slopes (Kruckeberg 1987). Huge avalanches of rock and snow released pent-up steam and gases resulting in a massive explosion on the north slope, repeated pyroclastic flows, ash fall-out and mudflows. No organisms survived in the pyroclastic blast zone, where a new landscape was created with 1200 °F (650 °C) pumice deposits more than 30 ft (10 m) deep in an area that was already thoroughly altered by the debris avalanche and blast. In this area, especially in the 'Pumice Plain,' a sterile landscape was created in which no organisms above or below ground survived. Survival was higher in the ash-fall zone, outside the blast zone, because ash descended with lower temperatures and less force. A friend was musing over this, imagining a gopher emerging at the end of winter to find its lovely green landscape reduced to a moonscape—'what the...?' In some areas the ground was still covered with snow, which protected plants, animals and mycorrhizae—plant root fungi—from heat. In intermediate disturbance areas, where ash deposits were less than 3 feet deep, many organisms survived in a variety of refugia, such as under snow and decaying logs, and within the soil. By the mid 1980s , plants were distributed across the ash zone, along with mycorrhizae and other decomposing fungi. Animals burrowed through the ash layer, helping to mix it with soil. Deep ash deposits have prevented colonization, including in previously forested areas.

Small populations of all kinds of plants, including forest trees, survived in protected microsites on the mountain. They have become important sources for recolonization of other areas. In the blowdown zone, caused by the blast of the eruption, plants like willows and alders recovered along stream areas within several years (Allen et al 2005). Not surprisingly, one of the most important plants to recolonize the mountain was the prairie lupine. As a nitrogen-fixer, it was able to colonize largely sterilized soils, and even help other plant species get established, due to the nutrients it contributes to the poor soils. Elk, birds, gophers and others gravitated to the patches of surviving plants, and played an enormous role in dispersing plant parts, seeds and mycorrhizae elsewhere within several years after the eruption.

Overall, a remarkably diverse assemblage of fungal, plant and animal species successfully dispersed through the disturbance zones within the first 25 years after the eruption. And yet, it will be centuries before the conifer-dominated forests typical of the Pacific Northwest re-occur; the mountain's ecology is still in a very early stage of succession (Crisafulli et al. 2005).

Not surprisingly, volcanic events can reduce diversity. Mount St. Helens had many eruptions prior to that of 1980: at least 15 eruptions are known from the last 5,000 years or so, with dormancy intervals ranging between 100 and 500 years (Kruckeberg 1987). Consequently, its pre-1980 flora was depauperate, or low in diversity, relative to neighboring volcanic mountains, and it lacked any endemic species. Recovery from its latest eruption is predicted to take hundreds to thousands of years.

So, if the Peaks erupted like this—it would have been an even more massive event, and a large portion of its biological communities could have been wiped out. A moist, maritime climate is regarded as very important in the ecological recovery of MSH. The Peaks collapsed and/or erupted during either a really warm interglacial phase or during a very cold glacial phase during the Pleistocene, based on the span of confidence intervals surrounding the ages of its lavas. Either climate would have led the ecology of the mountain in a very different direction, but regardless, cold-loving mountain species rule the mountain's forests and tundra today.

An interesting example of plant colonization of cinder cones involves the procession of ponderosa pine up the slopes of the very young Sunset Crater, a cinder cone which erupted around 930 years ago (Fig. 2.4). By

1966, trees as old as 430 years were found on its northern slopes and in its basin (Eggler 1966). So they are moving pretty fast, even on the edge of the arid Little Colorado River valley and Painted Desert; colonizing a cinder slope seems incredibly challenging in a number of ways, and weathering of cinders takes its time. Many cones in the SFVF have been colonized by grasslands and forests, especially to the west of the Peaks, where more precipitation occurs (Fig. 2.5).

The Sinagua Formation

"Steep volcanic mountains are generally blanketed on their lower slopes by a peripheral layer of weakly consolidated or poorly sorted sediment derived directly from their higher elevations" (Updike 1977). The Sinagua Formation encircles the Peaks, with its deposits resembling alluvial fans. It is the most extensive and youngest sedimentary formation in the area, covering more than 32,000 acres (13,000 ha) around the intermediate and lower slopes of the Peaks, with depths of more than 100 feet (30 m) in places (Fig. 2.8; Péwé & Updike 1976, Updike 1977). As the volcanic rocks of the Peaks have eroded, with great help from the Pleistocene, they have made this formation.

The Sinagua Formation was built up by countless debris flows—that's right, debris flows—carrying broken-down lava down all around the mountain (Fig. 2.8). A combination of slope instability, climate and water saturation precipitated the flows. Lots of water transformed the relatively immobile sediment into highly mobile slurries that moved across even relatively low gradients for long distances, such as along the pretty flat Deadman Wash all the way to the Little Colorado River. The presence of some rocks greater than 16 ft (5 m) in diameter in the debris fans around the base of the Peaks also suggests that these were mass movement flows; that would have been a sight to see. The most prominent debris fans are in Hart Prairie, Friedlein Prairie, east of Schultz Peak, and east and north of Sugarloaf (Holm 1988). Geomophologist Richard Hereford says that a debris flow moved all the way from Friedlein Prairie on the west side of the Peaks to Baderville and the Rio de Flag. Of course, many debris flows occurred as glaciers and snow melted on the Peaks at the end of the Pleistocene, providing rivers of water for such movement.

The Sinagua Formation is made up of six members, as each has distinctive origins and characteristics: Orion, Schultz Creek, Espil, Little Elden, Hochderffer and White Horse Members (Updike 1977). The Espil member comprises more than 90% of the Sinagua Formation; it alone encircles the entire mountain with multiple debris flow fans, which extend up into the canyons (Péwé & Updike 1976). The deposits are heterogeneous in size and and rock types, and consist of silt, sand, gravel, and boulders ranging up to 16 feet (4.9 m) in diameter, but ultimately unified in one formation. Some of the large boulders were carried more than 6 miles (9.5 km). Caliche and iron oxide locally cement the sediment. When drilled in several places, this formation was found to contain cinders, coarse sand, gravels and glacial outwash. Péwé & Updike (1976) suggest that some of the debris fans may have formed during the last several hundred thousand years.

Erosion and talus movement is occurring along bedrock exposures at all elevations, though particularly above treeline (Fig. 2.11). I remember hearing a rock fall from the upper slopes of Humphreys, which look as though they're made up of scree. Erosion is rapid enough that old logging roads have been nearly obliterated under talus.

A recent study of Schultz Creek sediment patterns determined that the Schultz Creek watershed is made of Pleistocene-aged dacite breccia and dacite lava domes with limited exposure of older basalt flows (Stempniewicz 2014). Prior to this study, the surface deposits along Schultz Creek were mapped cumulatively as undifferentiated alluvium, talus, till and colluvium (Holm 1988).

Soils

Though soils on the Peaks are in need of study, Gustav Pearson's early soils work provides a lot of insight (Pearson 1931). Of course, the soils on the mountain are derived entirely from its lavas. Pearson compared soils of the Peaks across an elevational gradient between Ft. Valley and the tundra, and found that they vary greatly as you go upslope. He found that the soils become more acidic with a climb in elevation, being almost neutral in ponderosa pine stands (pH 7.0) and finally rather acidic in the spruce-fir zone and above treeline (pH 6.5; Pearson 1931). Pearson suggested that less precipitation at lower elevations limits the leaching away of calcium carbonates and other alkaline salts, and their buildup might be the cause of less acidic soils there. Nitrogen levels were higher at higher elevations, with 0.10-0.12% N in soils in ponderosa pine stands, 0.26% in spruce-fir stands and 0.32% at timberline (Pearson 1931).

Pearson suggested that soil texture may be even more important to plants than chemistry in the dry Southwest, because of its influence on water absorption and retention. Soils in the ponderosa pine zone, did, on average, have a much higher percentage of clay than did soils upslope in mixed conifer and spruce-fir forests and at treeline. Higher elevation forests had soils comprised of sands, gravels and organic material. Ponderosa pine roots penetrate most deeply in sandy or gravelly soils, which also absorb water more readily than clays. Interestingly, the effects of drought are usually first seen in clay soils, because of their low permeability to both water and roots.

Water infiltrates more readily and more quickly in coarser soils, such as sandy, gravelly or stony soils in the higher forests; monsoon rains were found to penetrate down to 2 to 3 ft (to 1 m) in such soils. By contrast, 4 to 5 inches (0.1 m) of rain in July usually does not penetrate more than a foot in clay soil. Most clay is developed between 7,000 and 8,000 ft (2,100 to 2,400 m), due to temperature and moisture being sufficiently

2.11. Scree on steep slopes of Humphreys Peak.

abundant for the decomposition of minerals into clay.

On the Peaks, soils in the ponderosa pine zone receive the least moisture and have the highest temperatures. Ponderosa pine soils also have the least humus or organic debris: it's basically absent, and so never turns the soil brown. Humus influences the density of soil and contributes to moisture and nutrient retention.

Pearson summed it up like this: the ponderosa pine zone has denser, more clay-rich soils, with little humus or organic content, so it is harder for plants to absorb water there. By contrast, the mixed conifer and spruce-fir zone soils have a greater water-absorbing ability due to the presence of humus and because their coarser soils have a greater water-holding capacity.

Pearson (1931) found that tree species sent roots to different depths: ponderosa pine sent roots to depths of 6 ft (2 m), Douglas-fir sent roots to depths of 4 ft (~1.3 m), while those of Engelmann spruce rarely reached below 1 ft (~0.3 m) deep in the spruce-fir zone.

— SECTION 3 —

The Building of the Peaks' Ecological World

2.8 Ma to 12,600 years ago

The Pleistocene—Glacier Time

The Pleistocene Epoch, the world's latest ice age, lasted from about 2.6 Ma to 12,600 years ago (Table 1.1), so it was in full swing during most of the 2.8 million years that the mountain was building and then collapsing and/or erupting, and being colonized and recolonized by all kinds of plants and animals. This was the most recent global-scale climatic episode on Earth, and there are many signs of its occurrence throughout the world, including on the Peaks. There may have been as many as 50 glacial phases during the Pleistocene, and they accounted for more than 90% of this 2.6 million year-long epoch (Lisiecki & Roymo 2005). It got colder and then warmer and then colder, and on and on, but mostly colder through time.

I would love to have been around during the Pleistocene, to see massive glaciers displace mountain plants and animals, sending them down to lower elevations, to see the surreal megafauna, including fierce predators, roaming the continent (see below), and to see people on the move, crossing the Bering land bridge from Eurasia, following mammoths to the New World. What a time!

Our ability to understand the effects of the Pleistocene on the world's ecology has been greatly informed by the record of how plant communities responded to its climates. Scientists have detected such changes through analyzing and dating plant material, including pollen, that is preserved in sediment in lake bottoms and peat bogs, and in packrat middens. High in the Rockies, an understanding of the movement of plants during this time comes largely from peat bogs and lakes, since most high-elevation landscapes were scraped bare of vegetation and even soil by glaciers (Elias 2001). Such sources are relatively rare and mostly record conditions within the last 50,000 years, which includes the last glacial maximum. The various records reflect a world in motion. Plants were driven down in elevation and farther south, often ahead of massive continental glaciers.

Even changes in the ratios of oxygen isotopes, called Marine Isotope Stages (MIS), found in fossils deposited in ocean beds throughout the world confirm the occurrence of multiple Pleistocene glacial and interglacial phases: during glacial periods, the ratio of the heavier ^{18}O isotope in the oceans increased relative to that of ^{16}O, which was more readily bound up in glacial ice, and then the ratio switched during warmer times (Fig. 2.3; Cox & Moore 2005). Isotopes are slightly different versions of atoms, with ^{18}O having more neutrons than ^{16}O. This method has been used to determine climate patterns dating back several million years (Railsbacks et al. 2015)

So, as the mountain grew into its current form, the world impinging on it was in utter flux, with wild climate changes, glaciers perhaps cooling and eroding its lavas, their meltwater carrying sediment from the Inner Basin all the way to the Little Colorado River in debris flows and streams (Updike 1977, Ulrich 1984).

30,000 to 13,750 years ago

The Last Glacial Maximum

The last glacial phase of the Pleistocene, the Wisconsin, began about 85,000 years ago. By the *late* Wisconsin, also called the *last glacial maximum* (LGM, Table 1.1), the winter world was in full swing; cooler summers also occurred. Between about 30,000 and 13,750 years ago, the Laurentide ice sheet covered nearly all of Canada and much of the northern United States, and was up to *2 miles thick* (3.2 km) in places (Cox & Moore 2005). The Laurentide and neighboring Cordilleran ice sheets covered more than 6 million mi^2 (16 million km^2) and contained a third of the ice in the world's glaciers (Elias 2001). Glaciers covered many of the West's mountains, including the San Francisco Peaks and the White Mountains in Arizona, which were at or near the southern limit of glaciation in North America. Glaciers and ice caps extended down the south-running mountaintops of the Rockies and Sierra Nevada, capping mountains and filling mountain valleys with ice. A great portion of the San Juan Mountains in southern Colorado was covered with glaciers and ice sheets; they were enormous, covering nearly 1,930 mi^2 (5,000 km^2). The Animas Glacier extended downslope more than 40 miles (64 km) and was 100s of meters thick, forming moraines near Durango, Colorado, when it reached its maximum extent by about 26,000 years ago (Carrara 2011). The last major ice advance in the Rockies took place between about 30,000 and 13,750 years ago (Elias 2001).

By the middle of the Wisconsin's last glacial advance in the West, average summer temperatures were at least 13 °F cooler (Scott Anderson, personal communication 2018). Brrrrr. Of course, this made glaciers form in the Inner Basin and along some of the Peaks' outer flanks.

Glaciers on the Peaks

Today there are still signs of the Pleistocene's glaciers all over the world: rocks scarred and scoured by glacial movement, moraines or piles of rocks deposited by glaciers down valleys, and evidence of massive flooding as the glaciers melted: all very big stuff. In 1913, Henry Robinson described evidence of glaciers on the Peaks including moraines, an outwash plain in the lower Inner Basin and polished rock surfaces.

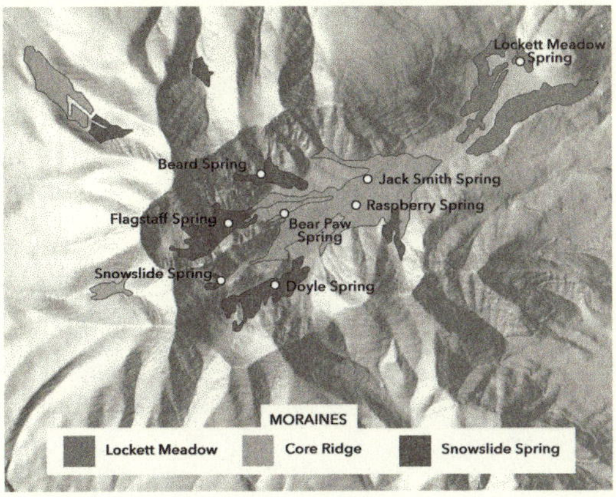

3.1. 1. Glacial moraines through the Inner Basin, San Francisco Peaks, with moraine-based springs. Lockett glaciation was the oldest, and Snowslide Spring the youngest, see Plate 2 for color. 2. Inner Basin moraines photograph. View to the northeast from the top of Agassiz Peak. Snowslide terminal moraine is in the right foreground and Core Ridge terminal moraines converge centerground, covered with light colored aspen stands. They are the undulating, rounded shapes. Stream action cut through the moraines, providing a route for the Inner Basin trail. Farther east, the valley is blocked by Sugarloaf Mountain. Photograph by Troy Péwé taken in 1968. Courtesy of Rick & Lisa Péwé.

A glacier is a large accumulation of snow transformed into ice as clear as ice cubes by pressure from the weight of overlying snow (MacDougall 2013). True glaciers must be permanent, persisting even through warmer seasons. Evidence of glaciers—after they've melted—includes sculpted cirques in the tops of mountains, which define their starting points, while glacial till or moraines define their extent down slope. *Cirques* or amphitheater-like headwalls in the upper slopes of mountains are where Pleistocene glaciers formed, carving out indentations in the steep sides of mountains. The largest

3.2. Cirques associated with glaciation between Humphreys and Agassiz Peaks. Imagine glaciers extending downslope from there. Courtesy of Ellen Wade.

cirques on the Peaks are in the western end of the Inner Basin, below Agassiz and Humphreys Peaks, separated by Core Ridge (Fig. 3.1, 3.2). The lower extent of a glacier ends in a terminal moraine or deposit of till, ranging from fine clay to large boulders (Cox & Moore 2005). What distinguishes glacial till from other rocky deposits, such as those left by flooding, is that the components of till are completely mixed rather than sorted by size. The degree of scarring or even polish on moraine rocks, and presence or absence of lichens, and of associated vegetation, have also been considered in determining the presence and age of moraines.

I have loved finally learning to see the moraines on the Peaks. It's kind of tricky, as they are so big-some hundreds of feet tall, and most are overgrown by forest and transected by stream erosion. You have to see them from above and a bit of a distance to get a sense of their shapes. The most pronounced are the Core Ridge moraines. They cluster about mid-Inner Basin, where the basin is the narrowest, and where a glacier in Doyle Canyon also fed into the basin (Fig. 3.1, Plate 2). They are often like softly-rounded domes or undulating lines, softened by aspens, which seem to dominate them (Fig. 3.1). The Lockett Glacier, the longest glacier in the Inner Basin, occurred from its cirques between 11,000 and 11,300 ft (3,350-4,000 m) and its terminal moraines occurring at about 8,500 feet (2,600 m) in Lockett Meadow (Fig. 3.1, Plate 2; Updike 1977).

Recent studies have determined that there were at least three glaciers in the Inner Basin, and each glacial event left a distinctive signature (Sharp 1942, Updike 1977, Poellot 2000). Cirques, enormous moraines, outwash areas, and glacial till were all studied. All studies noted the presence of considerable till and even a moraine along the slopes of Sugarloaf Mt. in the Inner Basin (Fig. 2.6, 3.3, Plate 2).

Although no dating techniques have been used to age the glacial remnants in the Inner Basin, the lavas of the Peaks' youngest volcano, Sugarloaf Mt., have been aged to around ~91,000 years old (Peters 2003). (They were previously dated at between 200,000 and 240,000 years old; Damon et al. 1974).

That date provides a very solid timeline for the maximum age of these three relatively young glaciers in the Inner Basin. With it, we know there were three glaciers in the Inner Basin within the last 91,000 years, because there is glacial debris from the *oldest* of the three plastered on the slopes of Sugarloaf (Fig. 2.6, 3.3, Plate 2; Peters 2003, Morgan et al. 2010, Updike 1977). This timeframe encompasses the last glacial maximum of the Pleistocene, between about 70,000 and 15,000 years ago. During this time the forest at Potato Lake, less than 50 miles (80 km) away and at about 7,000 feet (2,100 m) in elevation, was a closed spruce forest; this subalpine tree replaced ponderosa pine around the Peaks during the cold (Anderson et al. 2000). Brrr.

The presence of Sugarloaf in the basin makes this story pretty interesting, and ultimately suggests that there were also glaciers in the Inner Basin *before* as well as after its eruption. The eruption of Sugarloaf in the eastern mouth of the Inner Basin surely blocked further drainage to the east (Fig. 2.6). And yet, according to Updike (1977), there is also glacial outwash to the east of Sugarloaf all the way to Bonito Park, and farther east to where Sunset Crater now stands (Fig. 2.1; glacial deposits underlie Sunset Crater). Drilling records at Sunset Crater Monument show a 57 ft (17.4 m) thick deposit of outwash material there (Cosner 1962). These glacial deposits probably flowed out of the basin prior to

3.3. The lateral moraine deposited on the lower left side of Sugarloaf Mountain, during the Lockett Glaciation. Photo taken from the south ridge of Lockett Meadow.

the eruption of Sugarloaf, earlier in the Pleistocene. This sure would make for an interesting study! It's fun to imagine what an unimpeded flow of glacial outwash might have looked like with the eastern end of the Inner Basin being wide open.

Inner Basin moraines function as huge perched aquifers: in the midst of our evaporative world here, they store hundreds of millions of gallons of water—which is replenished in most years (Fig. 3.1; see Water section)! The moraines are also home to some of the most beautiful aspen groves on the mountain. Not surprisingly, most major Inner Basin springs issue from them.

Lockett Meadow Glacier was the oldest, largest known glacier in the Inner Basin (Fig. 3.1, Plate 2; Updike 1977, Poellot 2000). Cirques formed on the western peaks of the Inner Basin, and the ice from these coalesced, forming a glacier that extended all the way to Lockett Meadow, a distance of about 4.5 miles (7.2 km). In the middle of the Inner Basin, this glacier deposited nearly 650 feet (200 m) of till above the floor of the basin. It also left a long, high lateral moraine on the southern side of the Inner Basin that slopes down to Sugarloaf (Poellot 2000). Sugarloaf's western slopes are covered with outwash from the Lockett Meadow glaciation. A lateral moraine left by the Lockett Meadow Glacier can be seen on the north side of Sugarloaf (Fig. 3.3).

The second of the three glaciers, the **Core Ridge Glacier**, left the most visible and impressive glacial landforms in the Inner Basin (Fig. 3.1, Fig. 3.2, Plate 2). It extended down 2.6 miles (4.2 km) into the middle of the Inner Basin, where it deposited moraines up to 490 feet thick (149 m). The Core Ridge moraines form a substantial portion of the topography of the Inner Basin (Higgins 1998). The Core Ridge terminal moraine ends in

3.4. Aspens on a Core Ridge moraine in the Inner Basin.

3.5. Glacial outwash leading into Lockett Meadow.

the middle of the Inner Basin, just below the point where the municipal pipeline leaves the basin.

During the Core Ridge glaciation, glaciers also formed in canyons on the outer slopes of the Peaks, in Abineau and Rees canyons on the north side, and in Philomena and Snowbowl canyons on the west side. Outwash from the Snowbowl glacier reached the Rio do Flag (Updike 1977). Outwash from Abineau, Rees, and Philomena glaciers flowed into Deadman Wash. The ice age moraines in Snowbowl Canyon were altered by the development of Snowbowl ski area, so older maps were relied upon to determine its extent (Poellot 2000).

The youngest glacier, referred to as the **Snowslide Spring Glacier**, was the smallest, and its remains are the best preserved (Fig. 3.1, Plate 2). Evidence of this glacier's youth includes a lack of associated vegetation and lichens, and exposed rocks with little pitting or weathering (Poellot 2000). Its glaciers formed in 5 cirques in the western Inner Basin, but never coalesced. These glaciers are thought to have been about a mile long (1,600 m), and their moraines rise about 65 feet (20 m) above the basin floor. Several of these glaciers overtopped Core Ridge moraines, leaving their till there. There may have been several glacial advances within the Inner Basin during this event. Glaciers also formed in Abineau Canyon, and in Philomena Canyon on the west face of Humphreys Peak. During this glaciation, another major glacier formed in Doyle Canyon below Doyle Peak. The Inner Basin trail switchbacks across this moraine, and it supports one of the loveliest aspen groves in the Inner Basin (Fig. 3.4). These glaciers and their ability to generate soil through abrasive action created great habitats for aspen forests there.

Snowslide Spring glaciers may even have occurred after the Pleistocene, during a very cold time. There were glaciers in the Rockies during the Holocene, with the last having formed during the Little Ice Age, within the last several thousand years. Moraines from this time are the most common Holocene glacial deposits in the Front Range of the Rockies, where glaciers reached their greatest extent by the mid-1800s and have been retreating ever since (Elias 2001).

The flat, open floor of the Inner Basin is attributed to the erosional action of these glaciers. They made it more of a glacial basin than a fluvial or stream basin (Karátson et al. 2010).

Glacial outwash can be seen in the Inner Basin, and in several drainages below the mountain, from when the glaciers melted and water and debris flowed. In the Inner Basin above Lockett Meadow, the terrain is rocky with most rocks oriented 'downstream,' suggesting that they were rolling down in that direction along with glacial outwash (Fig. 3.5). Glacial outwash helped carve a 230 ft (70 m) deep drainage through volcanic ash and bedrock north of Sugarloaf and North Sugarloaf (Fig. 2.1). The outwash from this melting glacier carried volcanic debris from the Inner Basin all the way to the Little Colorado River via Deadman Wash (Updike 1977). What a sight that would have been, a stream of glacial melt moving rocks through the lowlands.

3.6. Telemark avalanche chute, running down the slope of Fremont Peak, with an enormous Scouler willow population at its base (see KachinaPeaks.org)!

Avalanches, glaciers and 'post-glacial' features

In 1973, the *Arizona Daily Sun*, the Flagstaff newspaper, reported that an extremely large snow avalanche ran from the upper southeastern slope of Humphreys Peak, shot over its base and ran into Dunnam Canyon, tearing out hundreds of large spruce and fir trees lining the sides of the existing avalanche zone (Dexter 1981). Interest in avalanches on the Peaks was sparked by this event.

Avalanches occur frequently on the Peaks: it is estimated that there have been as many as one thousand in the last 50 years (Dexter 1981, KachinaPeaks.org)! That's about 20 avalanches per year on average, crashing down more than 70 named avalanche paths (KachinaPeaks.org). Nearly all snow avalanches start on particularly steep slopes above treeline, between 33 and 40 degrees, in areas of abnormally high snow deposition, usually on leeward slopes. Consequently, paths in the Inner Basin run more frequently than paths on the outer flanks of the Peaks because that's where a lot of snow ends up, blowing over the western rim of the mountain (Dexter 1981). The single most active path is Telemark, in the north-facing canyon below Fremont Peak, which produced an average of 4.3 avalanches per year during Dexter's study. In summer you can see that there is also a rock avalanche chute here. Despite this, or perhaps because of it, the largest willow stand on the Peaks occurs at its base (Fig. 3.6).

Most heavy snow avalanches originate in the cirques that were carved and steepened by glaciers in the upper slopes of the Peaks. Leland Dexter even suggests that avalanches may have fed the glaciers. Both glaciers and avalanches thrive on wind-loaded snow.

Other paths that are particularly large and/or have frequent avalanche activity include Abineau Canyon, Beard Canyon,

3.7. Tree-shredding avalanche chute in Snowslide Canyon, June 2015.

3.8. Re-establishment of spruce and fir in Crossfire avalanche chute off of Humphreys Peak (see KachinaPeaks.org).

Dunnam Canyon, Snowslide Spring Canyon, North Fremont, East Fremont, Rees Peak, and Allison Clay Path (KachinaPeaks.org, Dexter 1981).

Despite the fact that the San Francisco Peaks is much smaller than the Rocky Mountains, its avalanche paths are as large (Leland Dexter, personal communication, 2015). Some of the avalanches on the Peaks are so powerful that they destroy full-grown trees in their path (Fig. 3.7, 3.8).

By measuring their tree rings, Dexter estimated that the tree-damaging events occurred approximately every 4-5 years. He often found that growth rings of trees that survived such damaging events, had greater growth rates, as competition from neighboring trees was eliminated (Dexter 1981).

In a way, the southwestern location of the Peaks predisposes it to avalanches. According to Dexter, Pacific storm systems with both cold air and high moisture account for most avalanche producing storms. Many times, snowfall from these storms is deposited over snowpack that is weakened by the abundant solar radiation that the Peaks receives, resulting in very large avalanches. Most avalanches occur during spring, when thawing begins.

Sometime after the arid mid-Holocene, a 'neoglacial' or cold phase occurred and numerous 'periglacial' or post-glacial features formed in the upper slopes of the Peaks and other western mountains (Péwé & Updike 1976). Periglacial features are basically mounds of rocks, formed in piles or lines of rocks piles, that form in mountains in response to freezing and thawing in really cold phases (Fig. 3.9). Their remnants are mostly found at more northerly latitudes and on more northerly exposures, which makes their occurrence at the low latitude of the Peaks interesting: the presence of these features strongly indicates the occurrence of colder,

3.9. Periglacial or post-glacial age features on the north side of Humphreys Peak, as stacked, wavy lines.

more rigorous climates during the late Holocene. Updike (1977) suggests that there are periglacial features on the Peaks that formed during very cold times during the last 4,000 years. Similar features have been reported from other regional low-latitude mountains, including the White Mountains, Kendrick Mountain, and Navajo Mountain (Updike 1977).

The 'rock waves' on the north slope of Humphreys (Fig. 3.9) provide an example of periglacial activity. This rock glacier is made of horizontal waves of rocks on the steep north slope of Humphreys Peak. It is comprised of a series of furrows and ridges occurring across the valley at the top of Abineau Canyon, and is described by Dexter as a product of seasonal freeze-thaw action upon waterlogged topsoils, which causes downslope movement of rock debris. Dexter regards this big feature as a pretty unique phenomenon.

The Pleistocene and the migration of plants

There were enormous migrations of plant populations throughout much of the world during the glacial and interglacial phases of the Pleistocene. This was also the case in the Southwest. Evidence of these migrations came with the discovery and study of plant parts preserved in sediment in lake bottoms, peat bogs and packrat middens....all over the world. These remains provide a record of the last 50,000 years, spanning most of the last glacial maximum, and they reveal a world in motion. Some of this record comes from Potato Lake and Walker Lake, near the base of the San Francisco Peaks (Anderson 1993, Anderson et al. 2000)!

Migrations

The migration of plants is hard to imagine. It's really plant *populations* that migrate: they shift their location in response to climate change. Imagine that plants are always dispersing their seeds around them, and if the conditions in a particular direction improve, i.e., away from an encroaching glacier, then their seeds that land in that direction will establish, spreading farther and farther out in that direction—establishing under the good conditions, but failing in the poor conditions. It was probably an incremental thing for the most part, one seed at a time becoming established and maturing to produce seeds on its own, and on and on. Imagine cold-loving Engelmann spruce seeds being blown from the Peaks down into ponderosa pine forests or even grasslands—as well as up onto glacial ice higher up on the mountain. When temperatures became very cold, spruce became established where pines died out. Some migrations during the Pleistocene were very fast. These rapid, massive migrations are a wonderful mystery, a phenomenon that we cannot entirely explain. While the seeds of some plants were moved along by animals, most were dependent on the wind.

Ice Age Forests on the move

The migration of boreal white spruce (*Picea glauca*) during the last glacial constitutes the ultimate displacement of forests by Pleistocene glaciers. During the Wisconsin, the Laurentide ice sheet spread across Canada and south of the Great Lakes, and obliterated the massive boreal forests—one of the world's most important carbon sinks—and scoured their habitats right down to bedrock. White spruce, a dominant species in Canada's boreal forests today, migrated south through much of the Plains States and Southeast in the United States by 18,000 years ago (Ritchie & MacDonald 1986). Mountain-dwelling limber pine and blue spruce, and even mountain-dwelling small mammals and land snails, were found in the plains during this time; today, little relict populations of them remain (Wells & Stewart 1987).

Glaciers occurred throughout the Rocky Mountains during the last glacial maximum, displacing both alpine tundra and mountain forests. These glacial phases were times of great range expansion for cold-loving, mountain plants, which of course have small ranges in warmer times like now, as they are confined to the upper, cooler spires of the mountains.

In response to the cold, mountain forests descended thousands of feet below current elevations and out laterally. According to Thomas Ledig et al. (2006), relatives of our Engelmann spruce, as well as fir and Douglas-fir, made it all the way to central Mexico, and more than once, during glacial phases (see below). These mountain trees were surely very widespread during the colder times, occurring even in the deserts of northern Mexico and the Intermountain West. It is estimated that more than a quarter of Arizona and New Mexico was covered by spruce during the last glacial maximum. I'll bet it was even more widespread than that! Spruce and fir forests covered extensive areas in the central and southern Rockies, the Mogollon Rim, the high plateaus of Utah, the Sierra Nevada and adjoining ranges of the Great Basin (Betancourt 2004).

So, bringing it back to the Peaks: The record from Potato Lake, 40 miles (64 km) southeast of the Peaks, provides one of the longest records of how the Peaks' plant communities were affected by the last glacial. This little lake (~4 acres, 1.6 ha) lies in a small depression in Coconino Sandstone at 7,290 ft (2,222 m). While Potato Lake is surrounded by endless ponderosa pine forest today just as the mountain is, things were very different there during the last glacial. NAU paleoecologist Scott Anderson (1993) took cores from the lake bottom's sediment. He found that the record of plant communities preserved in the muddy core extended back about 35,000 years, based on pollen and plant macrofossils that slowly collected over time, and the technique of radiocarbon dating some of them. This 30 ft (9 m) core of lake mud revealed remarkable changes in plant communities near the Peaks during the last glacial phase.

Between about 35,000 and 27,500 years ago, an open forest of Engelmann spruce, white fir, possibly limber pine and Douglas fir, with an understory of sagebrush, grasses and composites grew around Potato Lake, and surely around the Peaks, with much cooler temperatures than today (Anderson 1993, Anderson et al. 2000). Scott Anderson regards this forest as having been especially diverse during this time, resembling a mixed conifer forest, rather than a ponderosa pine forest. In fact, there was no evidence of ponderosa pine being present.

Ice Age Tundra on the move...

While major migrations of trees are hard enough to imagine, I find it even harder to imagine tundra plants on the move; in fact, singularly, this is the part of this whole story that never ceases to blow my mind. Tundra plants live above treeline, in the harshest conditions that mountains offer. They rarely exceed six inches in height and yet managed to migrate hundreds and thousands of miles in response to cold and glaciers. There are tundra plants on the Peaks that came all the way from the Arctic. These little plants are notoriously poor dispersers...migrating anywhere must have been a very slow process for these herbaceous perennials. As tundra plants were displaced by glaciers in the highest elevations of the Rockies during the Pleistocene, they moved downward and their ranges became much, much larger. In addition, tundra populations became far more interconnected, thanks to greater cold and moisture. Elias (2001) describes the archipelago of disjunct tundra populations in the Rockies today as having been a tundra *peninsula* during the late Pleistocene. It was surely because of these corridors that so many tundra plants made their way to the Peaks!

During the late Pleistocene, mountain sites below the glaciers were often dominated by tundra plants including *Bistorta*, *Pedicularis*, *Koenigia islandica*, and lots of members of the pink family (Caryophyllaceae), possibly including *Silene*, *Paronychia*, *Sagina*, *Minuarta* and *Melandrium*, and by sagebrush. Such assemblages formed at the edge of glaciers near Grand Teton and Yellowstone National Parks, and in the Rockies (Elias 2001).

Tundra to the Peaks... So just how cold would it have had to have been for tundra plants to make it through the Cold Desert to the Peaks? Average summer temperatures may have been at least 12-13 °F colder at the base of the Peaks during the late Pleistocene, making it plenty cold enough enough for tundra species to get here (Anderson, personal communication 2018). In considering routes of tundra plants to the south, certainly they migrated through the Rockies, but it may be that they also travelled along the western rim of the Colorado Plateau, which is a little higher in elevation than at its center (Fig. 3.10, Plate 3). Nobody knows what route they took from the Rockies, but if you look at a map of high elevation forests through the Colorado Plateau, there may have been several routes, considering that even today there is a near continuous ring of forests along this rim (Fig. 3.10). They came down from the Rockies through Utah and the Kaibab Plateau, and/or down through the Rockies to the White Mountains then along the Mogollon Rim to the Peaks. And then there are populations of some mountain species that also occur in the Chuska Mountains in northeastern Arizona, so maybe that was part of another route (Fig. 3.10, Plate 3).

We don't know when tundra plants arrived on the Peaks; there are no high elevation lakes or bogs to record their arrival. The Peaks was an evolving landform: it reached its tallest height within the last half million years, and then collapsed and/or erupted sometime between then and before Sugarloaf Mountain erupted 91,000 years ago (see above). There were a total of 4 major glacial phases during the last half million years of the Pleistocene, and the interglacial phases included 'megadroughts' (Fig. 2.3; Fawcett et al. 2011), so perhaps tundra species arrived several times and lost ground during drought, only to arrive again during the late Pleistocene. This onslaught of extreme climates would have challenged all of the mountain species, and especially those of the tundra. This is one of the more fascinating enigmas of our Pleistocene

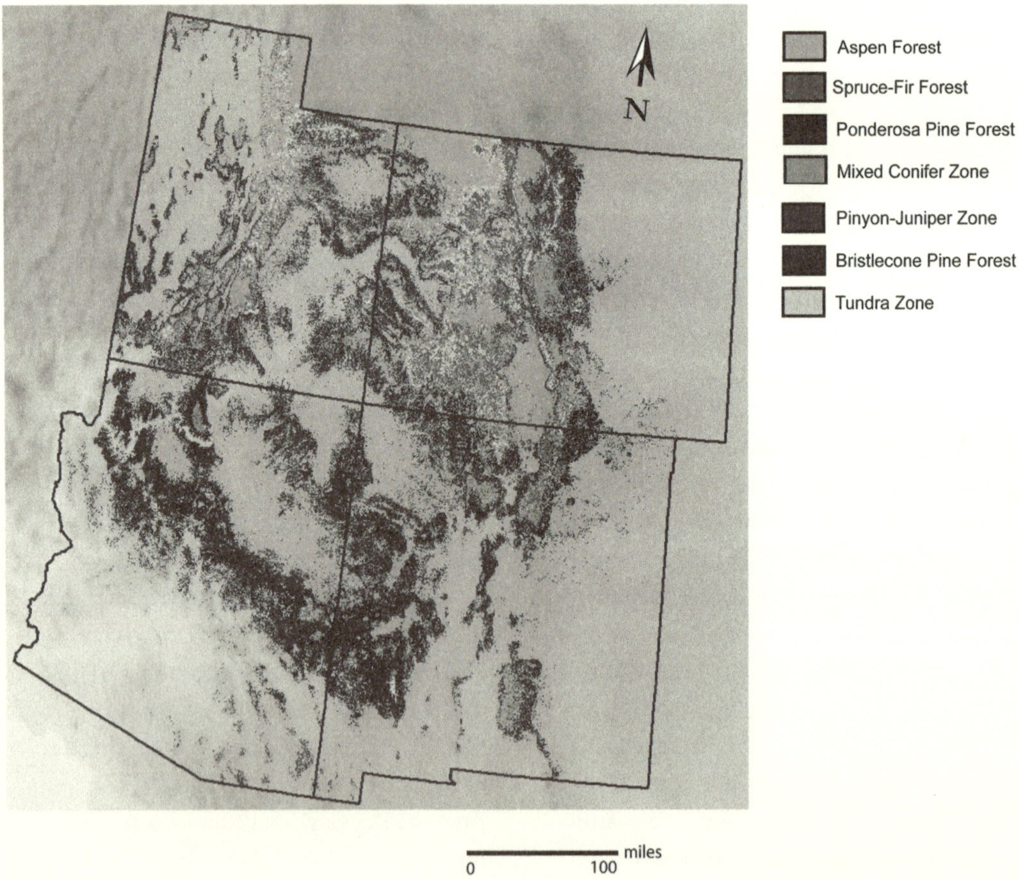

3.10. Colorado Plateau vegetation map. Note the Mogollon Rim, as the hefty southern edge of the Colorado Plateau. It runs diagonally across most of Arizona, dipping down just into New Mexico. Courtesy of John Vogel, DOI geographer and 2005 USGS GAP Data. See Plate 3.

history, imagining how these small plants ever made it to this sky island!

Less than 100 miles (160 km) to the north of the Peaks, Weng & Jackson (1999) describe the presence of tundra plants at 9,100 ft (2770 m) on the Kaibab Plateau on the north rim of Grand Canyon during the last glacial maximum, based on plant pollen from several lakes (Fracas and Bear Lakes). They also showed that alpine tundra dominated high elevation areas while spruce forests occurred below them. Interestingly, there are no tundra species there today. So, it may be that tundra plants reached the Peaks by migrating down the western edge of the Colorado Plateau, and then somehow managed to cross Grand Canyon and make it to north central Arizona, or....

A scientist at University of Arizona has an interesting strategy to learn about the arrival of tundra species on the Peaks; she intends to compare the genetic makeup of moss campion, *Silene acaulis* (Fig. 3.11, Plate 36), a very widespread tundra species in the Rockies and Utah, with an outlying population on the Peaks that represents its southernmost distribution. It will be an incredible study, as it may tell us when tundra species made it to the Peaks, how many times they came, and possibly what routes they took.

3.11. Moss campion, *Silene acaulis*, tundra plant on the San Francisco Peaks. Courtesy of Max Licher.

27,500 to 14,040 years ago
MIS 2 Glacial—last glacial maximum

White fir and Douglas fir were replaced by a closed Engelmann spruce forest, between about 27,500 and 14,000 years ago, at nearby Potato Lake; the glacial maximum—the coldest time—was reached. Large, northern glaciers correlate with lowered sea levels during this time. Average summer temperatures may have been at least 12-13 °F colder at the base of the Peaks during the late Pleistocene (Anderson, personal communication 2018). Today, the nearest Engelmann spruce forests are back on the Peaks! Subalpine forests are thought to have covered much of the Colorado Plateau above 7,150 ft (2,170 m) until the end of this last glacial phase (Betancourt 1984). There was still no evidence of ponderosa pine in the area.

So, if the base of the Peaks was colonized by dense spruce forest, you have to wonder what things were like on the mountain itself. There were extensive glaciers in the Inner Basin, and on some of the outer slopes of the Peaks, which limited the physical area colonized by plants. There are no vegetation records from the mountain at this time; such records from high elevations in mountains are rare everywhere, due to the erosional nature of steep slopes and the scouring action of glaciers.

The enigma of ponderosa pine during the last glacial maximum A major question about the Pleistocene in the West regards the whereabouts of ponderosa pine. Results from Potato Lake suggest that it showed up only at the end of the Pleistocene. So why was ponderosa pine absent during the last glacial and where was it? It seems to have been absent from much of the West. During the last glacial, ponderosa pine was restricted to south of 35° N, meaning south of Albuquerque NM and Flagstaff AZ (Betancourt 2004). There is a little evidence of it in southern Arizona and southern New Mexico during the late glacial (Anderson 1993, van Devender 1990). Ponderosa pine's pollen and plant parts are rather rare in packrat middens and lake core records during the late Pleistocene, in great contrast to its enormous presence today (Paklaian 2017).

During the full-glacial, the southwest generally had wetter winters and cooler summers with less monsoonal precipitation than today. During this time the jet stream moved south, which brought Pacific air masses across the Southwest, resulting in more precipitation generally, but less monsoonal storm action during the summer. It has been suggested that this reduced monsoon may have contributed to the apparent lack of ponderosa pine in the region during the late Pleistocene (Vankat 2013). Recent studies have determined that young ponderosa pines are dependent on monsoon moisture to establish and grow (Kerhoulas et al. 2013b). A later return of monsoons and with them, increased lightning-caused fires, may have contributed to this pine's rapid return to the region.

Down into the deserts

Even the distributions of desert plants were reduced during the Wisconsin cold, with many found in low-elevation refugia, such as areas below 1,000 ft (300 m) in Death Valley and along the Gulf of California (Betancourt 2004). Sonoran Desert species such as saguaros and foothills palo verde may have been reduced to small areas in Sonora (Van Devender 1990). Pinyon-juniper woodlands covered great expanses of what is today covered with desert scrub, between 1,000 and 5,500 ft (300-1,680 m). Mixed conifer forests and woodlands dominated rocky terrain above 4,500 ft (1,370 m) on the Colorado Plateau and in the Great Basin, where today there are ~50 million acres (20 million ha) of pinyon-juniper woodlands (Betancourt 2004).

One of the most important records of climate and plants in the West comes from Utah, from the Great Salt Lake, about 500 miles (805 km) north of the Peaks. The lake cores analyzed in this study present a record of plant migrations extending back into the Miocene, more than 13 Ma, based on pollen found in sediment deposits in the lake (Davis & Moutoux 1998). While this study describes the emerging cold deserts of western North America since the late Miocene—in response to the rain shadows created by the uplift of the Sierra Nevada (Axelrod 1990)—it also shows changes in plant communities in response to glacial cycles during the Pleistocene.

According to Davis & Moutoux (1998), desert vegetation near Great Salt Lake, including plants such as jointfir (*Ephedra*) and the chenopod greasewood (*Sarcobatus*) increased in prominence during the warm, interglacial periods of the early Pleistocene, while sagebrush was prominent during glacial phases of the early Pleistocene. During the last glacial maximum of the late Wisconsin, the expansion of desert plant communities ended abruptly, and pines, junipers, greasewood and grasses were replaced by spruce, fir, Douglas-fir and woody sage. Amazing to imagine spruce and fir replacing desert plants: now that's climate change driving huge biological changes.

Pleistocene Animals

Paleoindians of the Clovis culture and a number of their favorite prey species, including mammoths, were moving through Arizona and the rest of North America by 13,000 years ago. They hailed from the people who came across the Bering land bridge from Asia to the New World during the late Pleistocene, along a corridor that was exposed as ocean levels lowered with glaciation. That sure doesn't seem like all that long ago to me. The first evidence of these early people visiting Arizona appeared in southeast Arizona in the 1930s with the discovery of a Clovis point, and since then more than 100 points made by these people have been found through the state. About 25 points have been found in northern Arizona, and 20 of those have been found in every direction around the Peaks, with one found 18 miles (29 km) away (Peter Pilles and Phillip Geib, personal communication 2017). Their discovery provides evidence of early people hunting large prey near the Peaks. Mammoth and bison also came across the Bering Straits from Asia.

During the last glacial maximum, there was a remarkable diversity and abundance of large grazing and browsing mammals, including types of horse, camel, bison, ground sloth, mountain goat, bighorn sheep, mammoth, mastodon, llama, pronghorn, muskoxen, tapir and elk. They are referred to as the Pleistocene megafauna. Their remains have been found north and south of the Peaks, as well as through much of North America (Table 3.1). In southern Arizona, their fossilized bones and teeth have been found in sedimentary rocks, while in northern Arizona and farther north, evidence for them comes mostly from caves that have protected their dung or excrement, and various body parts. It all depends on the rocks, it seems.

Mastodon, which had been here quite awhile, apparently fed preferentially on spruce foliage...hard to imagine (Table 3.1; Metcalfe et al. 2013). They must have been well-fed during the Pleistocene, since spruce was often very widespread! Perhaps the extensive spruce forests around the Peaks drew them here; its nice to imagine mastodon browsing spruce on the Peaks. Mammoth remains and Paleoindian sites are well-represented in northern Arizona, often occurring along waterways and river valleys.

What a world it must have been. Doug Peacock (2013) really brings it to life in his book on the late Pleistocene, in which he describes people having to contend not only with the journey across the Bering Straits into an icy, unknown world, but also with phenomenal predators (Table 3.1), including sabertooth cats—also called 'stabbing cats', short-faced bears, and the American lion, which was up to 8 feet long (2.4 m), not including its tail. The short-faced bear was as tall as a man—when standing on all four legs! Peacock marvels at the courage and tenacity it must have taken to face and survive such voracious predators on top of everything else. The closest fossils of these fierce creatures are known from sites in Utah, California and Arizona; in Arizona, short-faced bear fossils were found about 70 miles (113 km) north of the Peaks, in a cave in Grand Canyon.

By the end of the Pleistocene, most of these animals had gone extinct. Horses migrated to Asia and South America and went extinct in North America, only to return with the Spanish in the 1400s and 1500s. Bison and the pronghorn survived and are still around today. A debate continues as to why most of the megafauna went extinct, including arguments for climate change and overhunting by people.

The Clovis people, one of the earliest North American human cultures, spread across North America within several hundred years. So, it was a time of great movement.

Table 3.1. A partial list of the Pleistocene megafauna fossils—denoted by X—found on the Colorado Plateau, in southern Arizona, and at the La Brea Tar Pits, CA.

Taxon	Plateau dung[1]	Southern Arizona fossils[2] Early*	Southern Arizona fossils[2] Late Pleistocene**	La Brea Tar Pits CA[3]
HERBIVORES				
Horse	X	X	X	X
Camel	X	X	X	X
Bison	X	--	X	X
Ground sloth	X	X	X	X
Mtn. Goat	X	--	--	--
Bighorn sheep	X	--	--	X
Mammoth	X	X°	X°	X
Mastodon	--	--	X	X
Llama	--	X	--	--
Pronghorn	--	X	X	X
Musk ox	--	--	X	--
Tapir	--	--	X	X
Elk	--	--	X	--
OMNIVORES				
Humans	--	--	X	--
PREDATORS				
Saber-tooth cat	--	X	--	X
Short-faced bear	--	X	--	X
Dire wolf	--	--	X	X
Pleistocene jaguar	--	--	X	--
American lion	--	--	--	X

1 = Mead & Agenbroad 1992
2 = Lindsay 1978
3 = La Brea website

* = early Pleistocene = 1.8 Ma
** = late Pleistocene = 0.5 Ma
°= uncommon, then common in late Pleistocene.

—— 11,700 years ago to the present —— MIS 1 Interglacial

The Holocene
The Peaks Becomes A Sky Island

The Holocene started about 11,700 years ago. It may simply be an interglacial phase of the Pleistocene; so stay tuned. In any event, it is the warm period since the end of the Pleistocene's last cold glacial. This warming brought extraordinary changes in the distributions and the evolution of all kinds of species. Life was on the move again.

Deglaciation or the melting away of glaciers in the West occurred between about 15,000 and 12,000 years ago (Betancourt 2004). It appears that it took less time for glaciers to melt than to form (Fig. 2.3; Lomolino et al. 2006). Deglaciation was accelerated by increased atmospheric dust, water vapor and especially greenhouse gases such as carbon dioxide and methane. Rapid global warming must have driven many local populations extinct: just a little too fast for either adaptation or emigration.

The warming led to spectacular flooding, as glaciers melted all over the West and much of the world (see Waring 2011). Evidence of this for the Peaks includes its glacial debris reaching the Little Colorado River.

Across northern Arizona there is evidence of a major period of downcutting of drainages and sediment banks along streams and rivers, perhaps due to increased runoff from melting glaciers (Joyal 2004, Anderson et al. 2000, Neff et al. 2003, Hereford 1984). This flood-induced erosion would end between 8,000 and 6,000 years ago, with the onset of the dry mid-Holocene (see below). Studies suggest that drainages near the Peaks flushed or flooded in the early Holocene. Flows from the Peaks may have occurred by 8,000 years ago, and indeed, Victoria Stempniewicz's earliest accumulation date for Schultz Creek dates to ~6,500 years ago (Stempniewicz 2014, Joyal 2004), suggesting that things had been eroding rather than accumulating previously.

As warming continued, the San Francisco Peaks would become a sky island, and surely not for the first time. As mentioned above, several interglacial phases, with 'mega-drought' climates, occurred following the final volcanic building of the mountain (Fig. 2.3, Table 1.1). This would have isolated mountain species that had already arrived, or perhaps even driven them to local extinction. But more about that later.

Along with warming temperatures, southwestern monsoon storms returned and may have been more intense than they are today (Carrara 2011). So it may have been a time of great productivity in plant growth and expansion in the Southwest. Perhaps this luscious weather allowed ponderosa pine to expand into its current large range as quickly as it has, travelling thousands of miles within the last 10,000 years.

Migrations in reverse

I thought the story of Holocene warming effects around here would be easier, or less complicated than that of the glacial phases. But in fact, the migrations of plants and animals 'home' as climates warmed have their own strange stories. As glaciers receded, plants began returning to sites where they'd occurred before the last glacial phase. In some western mountains, recolonization was slow because intense scouring by glaciers took landscapes down to bedrock. And then there is a certain randomness to many distributions, with Engelmann spruce occurring on Kendrick Mountain and the Peaks but not on Bill Williams Mountain (Fig. 2.1).

I can't quite shake the notion of plants adjusting to the end of the last ice age resembling troops returning from battle, but, of course, that isn't it at all. The reality is that cold times are good times for mountain and boreal species, while warmer times are good for lower elevation species, when each lives fully, spreading and invading wildly.

What migrations might have looked like

The rapid migration of species such as ponderosa pine in warming climates may also have been helped out by drought, fire and even insect outbreaks in spruce and fir forests that had expanded into lower elevations during the cold. There is evidence, at this time, for stand-replacing fires in the West, including on the Kaibab Plateau and in Schultz Creek at the base of the Peaks (Betancourt 2004, Weng & Jackson 1999, Stempniewicz 2014). After spruce and fir reached large numbers in this area, charcoal accumulation rates increased rapidly by nearly 13,500 years ago.

It's easy to imagine birds facilitating many plant migrations, such as those of juniper, and bird pines such as limber pine, pinyon and white bark pine; these plants have coevolved with corvids such as jays, which are quite good at moving their seeds. But the dominant mountain conifers of the West were on their own, at the mercy of wind and time. Species like spruce and fir have no special adaptations for dispersal, and are known to take hundreds of years to recolonize even nearby burn sites (e.g., Lanner & Vander Wall 1980); so their migrating hundreds and thousands of miles kind of defies the imagination. Today, the mountain island floras of the Great Basin are dominated by species with no apparent adaptations for long-range dispersal, suggesting that many of them got there one step at a time from the Rockies (Harper et al. 1978).

As the glaciers receded in the southern Rockies, forests familiar to us returned to the mountains (Carrara 2011). Spruce migrated back to subalpine habitats in the Rockies by 12,300 years ago, after an absence of nearly 18,000 years! Elias (2001) also found evidence of montane and subalpine insect species returning to high elevation sites. At lower elevations, spruce and fir populations declined by about 11,000 years ago, and ponderosa pine moved in with warming. At Bear Lake on the Kaibab Plateau, ponderosa pine was back within a thousand years of when spruce and fir populations began to decline (Weng & Jackson 1999).

12,400 years ago on the Peaks

Modern plant communities were present in the vicinity of the Peaks by about 12,400 years ago. Spruce, fir, limber pine, aspen and other species climbed back up into the mountain to make up its high elevation forests, as the glaciers there melted away. Ponderosa pine was already common around the mountain by then (Anderson et al. 2000). Oaks, grasses and sunflowers also became common, suggesting open forests, while spruce, fir, sagebrush, and pines such as limber pine had all moved on. Ponderosa pine exploded on the scene in the Southwest by 11,000 years ago (Betancourt, Lyford & Jackson, unpublished data, Betancourt 1990, Paklaian 2017).

The movement of ponderosa pine into its now enormous distribution, after being absent from this region during the last glacial maximum, is one of the most intriguing stories of all. It is thought that ponderosa pine migrated into its current enormous western range from southern refugia during the last 11,000 years. After watching a banner year for ponderosa pine cone production combined with extended summer precipitation in northern Arizona in 2013, and the resulting seedling explosion of this species, I can imagine how such a rapid migration/expansion could happen (Fig. 3.12). Multiply years like 2013 by several hundred or several thousand years, and ponderosa pine might move along pretty fast. The climate at the end of the Pleistocene was warmer and also very wet during summer, which would provide excellent conditions for this species to thrive for many reasons, including the fact that it naturally germinates with summer rains.

Another measure of warming conditions was seen in a shift from cold-water adapted aquatic organisms to more temperate forms in nearby Walnut Canyon approximately 10,800 years ago (Hevly & Karlstrom 1974).

9,000 to 8,000 years ago

People at Bismark Lake

The earliest confirmed date for human presence on the Peaks is between 9,000 and 8,000 years ago, during the early Archaic period. The people who left these traces were closely related to those who came across the Bering land bridge. Late Paleoindians left a piece of an Eden-style projectile point at Bismarck Lake, perhaps at a seasonal camp, on the western slope of the mountain (Fig. 3.13; Peter Pilles, personal communication 2017). This projectile point, made of nonlocal chert, was projected or thrown with an atlatl, right along the shoreline of Bismarck Lake!

3.13. A late Paleoindian projectile point from Bismarck Lake, on the west slopes of the Peaks. Courtesy of Peter Pilles & USFS.

3.12. Cluster of large ponderosa pine seedlings establishing after fire.

8,500 and 2,000 years ago

Middle Holocene—an arid time

Climate in this area and in other parts of the world is thought to have been very warm and dry between about 8,500 and 2,000 years ago; it is often referred to as the Holocene Aridness, and was so dry at times that some regions are devoid of any record of water or vegetation. Arid conditions are suggested at this time for Schultz Creek at the base of the Peaks, in lakes around the mountain, and on the North Rim. A severe drop in lake levels occurred in Potato Lake, 40 miles (64 km) from the Peaks; this lake that had told so much about local vegetation changes during the Pleistocene either had very low lake levels or went dry during the mid-Holocene (Anderson 1993). It was so dry that there is no record of any deposition of organic material or sediment, indicating that the lake dried out frequently. During this time, Walker Lake, on the north side of the mountain, reached its lowest level in 20,000 years (Hevly 1985). Sediment cores from Stoneman Lake to the south show that amounts of pollen and sediment influx decreased, also indicating drier times, between 4500 and 2500 years ago (Harsbargen 1994).

Schultz Creek, which drains much of the south side of the Peaks, provides a record of mid-Holocene warmth and dryness: By around 6,500 years ago, in the absence of big storms and floods, sediment began to accumulate in Schultz Creek, and continued to do so slowly and steadily, up until about one hundred years ago (Fig 2.6, 3.14; Stempniewicz 2014). Fires spurred by dryness may have contributed to erosion of soil and debris from the steep slopes of the Peaks, leading to further aggradation, as occurred in drainages closer to the Mogollon Rim (Joyal 2004).

Mountain forests in the West responded to mid-Holocene warming by climbing higher. In the southern Rockies, treeline populations of spruce and fir migrated between 300 and 1,000 ft (90-300 m) above where they grow today, with average summer temperatures possibly being 5-6 °F higher (Carrara 2011, Scott Anderson personal communication 2018). Similarly, the remains of Great Basin bristlecone pine (*Pinus longaeva*) dating back to the mid-Holocene are found at elevations 500 ft (150 m) over current treeline populations in the White Mountains in California (LaMarche 1973). Carrara (2011) cites over 20 studies of elevated treelines in western mountains during the mid-Holocene.

Several forest species occur at higher elevations on the Peaks today than they did 140 years ago (Cocke et al. 2005), so perhaps they also climbed higher during the mid-Holocene. Perhaps subalpine forests and

3.14. An arroyo willow, *Salix lasiolepis*, stand along Schultz Creek.

tundra plants were crowded together on the upper slopes of the Peaks, with mixed conifer species and ponderosa pines on their heels.

Late Holocene on the San Francisco Peaks, the last 2,000 years...

The record of climate, fire and the forests on the Peaks really picks up in the last several thousand years, with studies of the tree rings and scars from fires in its forests. By about 2,000 years ago, the dry conditions were replaced with a time of unparalleled climatic variability (Vankat 2011). Some really great studies describe conditions of climate, the forests, and fire on the Peaks through this time (Merriam 1890, Leiberg 1904, Pearson 1931, Hevly 1985, Cocke et al. 2005, Heinlein et al. 2005, Salzer & Kipfmueller 2005, Margolis et al. 2011).

Two periods stand out in the climate record of the last thousand years: a warm, dry spell called the Medieval Climate Anomaly Period (MCA) between 1,000 and 700 years ago, followed by a colder interval, the Little Ice Age, in the last 500 years (MacDougal 2013). In Europe, the MCA was generally characterized as a time of relatively warm, benign climate, with many stretches of time that were highly favorable for agriculture. During the Little Ice Age, it was cold enough and wet enough for the formation of glaciers, as in the Front Range of the Rockies (see glaciers above).

Salzer & Kipfmueller's study, however, found that climate on the Peaks and nearby was highly variable, lacking evidence of the sustained warmth or cold predicted for these episodes. They found that there were 30 extremely cool and 26 extremely warm periods, and 30 extreme wet and 35 extreme dry periods on the Peaks during the last the last several thousand years (Salzer & Kipfmueller 2005). The driest interval was from 1818 to 1823, while the longest wet period was between 1905 and 1922 (known as the Early 20th Century Pluvial). The ecological consequences of the wet period were well-documented and enormous (see Pearson below).

The last several hundred years have been especially dynamic climatically, and with many other factors converging and quickly. EuroAmericans started to have an enormous impact on the mountain, in studying it, but also by unleashing herds of cattle and sheep and allowing them to graze mercilessly starting around the 1860s, by overlogging many of the ponderosa pine stands surrounding the mountain, and by suppressing fires in the late 1800s.

The Evolution Of Plants and Animals

The wild climates of the Pleistocene and the Holocene may have led to significant evolutionary changes in many plants and animals. Sometimes glaciers drove mountain populations downward into glacier-free, isolated refugia. Some species that live on the Peaks, including limber pine (*Pinus flexilis*), and a raptor, the Northern Goshawk (*Accipiter gentilis*, Fig. 4.23), may have occurred in ice-age refugia near the Peaks and Mogollon Rim, forming distinctive genetic populations in the process of being isolated from other populations and exposed to local natural selection. They have since migrated outward into the West, with warming (Mitton et al. 2000, Bayard de Volo et al. 2013).

Nearly all of the conifer species on the Peaks have diverged genetically during Holocene warming, resulting in clines of genetic variation from their arid south to more mesic northern populations (Goodrich et al. 2016, Hamrick 1976, Rehfeldt 1990, 1993, 2004, Rehfeldt et al. 2014). A cline represents a gradation of variation in one or more traits. In ponderosa pine forests extending from southern Arizona to the Rockies, Holocene warming has favored variation of important traits along the elevational climb from Arizona to the Rockies, with greatest genetic differences occurring between populations in Arizona and New Mexico and those in Utah and Colorado (Rehfeldt 1990; see Ponderosa pine and other conifer clines below).

Frankly, I wonder if the poorly-studied corkbark fir on the Peaks might have arisen at the end of the Pleistocene (Fig. 3.15). Corkbark fir represents a southern extension of the widespread subalpine fir, which has very thin bark—like other subalpine trees. By contrast, corkbark fir has very thick bark. Corkbark occurs only in the southern Rockies and in Arizona and New Mexico, and its thicker bark might help it to tolerate the warmer and drier conditions in the Southwest. This is another study that needs to be done.

The North American genus, *Penstemon*, is thought to have originated in the Rocky Mountains, and radiated and diversified rapidly during the Pleistocene. The center of its diversity lies in the Intermountain West, and today this genus is comprised of more than 300 hundred species, subspecies and varieties with an amazing ecology (Wolfe et al. 2002, 2006, Waring 2011 for a review). A combination of the topographic and geological complexity of this region, along with glacial-interglacial disturbances and even interactions with pollinators is thought to have led to the remarkably rapid evolution and radiation of penstemons. Five species of penstemons occur between the ponderosa pine forests and the tundra on the Peaks. The lovely *Penstemon whippleanus* occurs above treeline there (Fig. 3.16, Plate 33). The late James Reveal, esteemed Polygonaceae systematist, suggested that the highly diverse buckwheat genus, *Eriogonum*, radiated rapidly with the expansion of southwestern deserts at the end of the Pleistocene (Reveal 1979, Eflora.org).

Also, intriguingly, the Abert squirrel (*Sciurus aberti*), which is tightly associated with ponderosa pine, is comprised of six subspecies, some of which only originated in the late Pleistocene. Lamb et al (1997) suggest that the genetic makeup of this species was strongly and perhaps repeatedly influenced by the many climate changes associated with the Pleistocene (Fig. 3.17). This also applies to the Kaibab squirrel (*S. a. kaibabensis*) on the north rim of Grand Canyon, which is thought to have originated in the late Pleistocene. I

3.15. Cork bark fir, *Abies lasiocarpa* var. *arizonica*. Note its light-colored, corky stem.

can't help but imagine ponderosa migrating/expanding north very fast and in enormous populations with this squirrel hanging on for dear life—although perhaps the migration was slow enough that it wasn't really obvious.. kind of like today. And then with warming, both pines and squirrels became isolated and diversified.

Migrating from the fringes

A relative of Engelmann spruce migrated all the way to Mexico during the Pleistocene and now faces extinction there with warming. It provides an excellent example of what many species are facing as global warming heats the planet. With warming, spruce populations there have shrunk, leaving only a few isolated populations perched on mountaintops in the Southwest and northern Mexico. Fossil remnants of spruce can be found at sites in Mexico where living members of the species no longer occur. As spruce populations

3.16. Whipple's penstemon, *Penstemon whippleanus*, in the tundra. Courtesy of Max Licher.

3.17. Abert's squirrel, *Sciurus abertii*, on ponderosa pine. Courtesy of Gary Botello.

have become isolated they no longer receive pollen from other populations, so good old healthy outcrossing or outbreeding ended. These small populations took to inbreeding, which has been disastrous, resulting in the production of seeds that are not viable, so successful reproduction is pretty much over. And yet…there are a few glimmers of hope, as in the evolution of *Picea chihuahuana*, a Mexican spruce that is more tolerant of heat than other spruce species (Ledig et al. 2004)!

Human Presence On The Mountain—The Last 2,000 Years

Native Americans

There are hundreds of prehistoric Native American sites on the San Francisco Peaks, dating from the Paleoindian period (9500-6500 B.C., see above), and continuing through the Archaic Period (B.C. 6500 to A.D. 600), according to Coconino National Forest Archaeologists Peter Pilles and Jeremy Haines. After 600 A.D., more evidence of agriculture was seen along with occupation by two different groups descendant from the earlier Archaic period—the Cohonina and Sinagua cultures. Cohonina were dominant to the west and northwest of the Peaks, while the Sinagua were especially numerous on the eastern flanks of the mountain and farther east, almost to Canyon Diablo.

Cohonina and Sinagua sites occur on all sides of the mountain, and especially on the eastern flanks, perhaps because it was warm enough there to support extensive agricultural efforts. Besides farming in the alluvial park lands, such as Doney Park, Black Bill, Lenox and Bonito Parks, they also constructed rock check dams along the washes that flow east off of Mount Elden to create small, scattered farm plots. Camp sites near springs were especially popular, as evidenced by the numerous lithics scatters found at so many of them; they probably represented hunting camps. Most camps were below 7,500 feet (2,300 m) in elevation, though there are a few lithics scatters in the Inner Basin.

Most sites consist of small masonry rooms used seasonally while growing crops. Sinagua habitation sites typically consisted of small clusters of pit houses.

There is evidence of Sinagua Indians having lived along the Rio de Flag, in the vicinity of today's Museum of Northern Arizona, between 900 and 1100 A.D. Dwellings included clusters of pit houses, suggesting residency, though they were probably seasonal (Peter Pilles, personal communication, 2016). Some pit houses found during this period were unusually large, and thought to have been used for community or ceremonial purposes. There were also field houses, which supported agricultural efforts. There is evidence of check dams across washes to capture water-deposited sediment to create farm plots. There was considerable trade between the Sinagua here and the Cohonina Indians to the north (Pilles 1979). The Cohonina lived on the north side of the Peaks, all the way to Grand Canyon. The Sinagua Indians were more agriculturally inclined than the Cohonina. There is also evidence of Sinagua and Cohonina presence along Schultz Creek. It might have been quite mesic or wet then, before the water in so many of the springs was diverted.

With the eruption of Sunset Crater about 930 years ago, Sinaguan populations expanded northward into the Wupatki area and farther east (Fig. 2.1). In places, 50 to 80 sites per 250 acres (100 ha) have been found (Pilles 1979). Sinaguan culture blossomed between 1150 and 1250 A.D. (Pilles 1979). Elden Pueblo was built and occupied during this time. It lies at the base of Mount Elden, just southeast of the Peaks and may have housed 150 people. Most Sinagua left the Flagstaff area after about 1275 to live on Anderson Mesa. They stayed there for a century or so before continuing eastward to the Little Colorado River valley, where they contributed to the blending of various prehistoric groups that would eventually become Hopi (Pilles 1979).

The San Francisco Peaks is sacred to many people. Native American tribes who revere the mountain include the Navajo, Hopi, Zuni, Tewa, Hualapai, Havasupai, Yavapai-Apache, Yavapai-Prescott, Tonto Apache, White Mountain Apache, San Carlos Apache, San Juan Southern Paiute, Fort McDowell Mohave Apache and the Acoma. All tribes hold that disturbance of the mountain violates their way of life and traditions. According to Peter Pilles, most mountains in the world are revered by native people. In part, this reflects gratitude for the resources, including water, plants and animals, that mountains provide to local people. Former Hopi chairman Vernon Massayesva wrote that the Peaks is "the shrine we look to because it is the home of ancient Katsina spirits, emissaries of life…Sometimes we felt we could touch the mountain near Flagstaff" (Gulliford 2000). Though sixty-five miles to the southwest, the Peaks are clearly visible from the Hopi village of Oraibi. Members of the Hopi tribe make annual pilgrimages to the Peaks to leave offerings for the Katsinas who live there, and gather spruce branches for dances and ceremonies.

Havasupai Indians call the Peaks Wii Hagnbaja or Wii Hlaa', meaning 'Snowy Mountain' or 'Moon Mountain', the latter being its ceremonial name (Hirst 2014). They believe that the Peaks drive the precipitation cycle, providing water for all living things. The Peaks links them to heaven. Creation of all living beings occurs there.

The Hopi call the Peaks Nuvatukya'ovi meaning 'Mountaintop Home of Snow'. To the Hopi, the Peaks is the home of the katsinam (Hirst 2014). The Hopi summon these ancestral spirits to their villages each year beginning with the Bean Ceremony in February, and bid them farewell with the Homecoming Ceremony when the katsinam send the summer rains in July.

The Navajo peoples name for Peaks is 'Doko'oo'sliid', meaning 'Gleaming Summit'. The Peaks, Mt. Blanca and Mt. Hesperus in Colorado, and Mt. Taylor in Mew Mexico, create an area within which the Navajo find harmony (Hirst 2014).

The Zuni Indians call the Peaks Sunha:kwin K'ya:bachu Yalanne, meaning 'Highest Mountain to the West'. Like the Hopi, the Zuni center their spiritual beings on the Peaks, from which they send rain and snow.

For as long as anyone can remember, Hopi Indians have hunted in this area and carried Douglas-fir poles from the San Francisco Peaks to the Hopi mesas for the construction of kivas (Peter Pilles, personal communication, 2017). Many native people hunted and foraged on the Peaks.

Later, Franciscan missionaries establishing themselves at Hopi also used Douglas-fir poles from the mountain in the construction of their churches there. In 1629, the Peaks was named in honor of Saint Francis of Assisi, by friars who established a mission at the Hopi Indian village of Oraibi on the Hopi Mesas (Cline 1982).

Euro-American explorers

Between 1851 and 1878, EuroAmerican explorers began moving through the area near the Peaks. The first official expedition through the region was the Sitgreaves party, in 1851, with a great cast of characters, including Antoine Leroux, a renowned guide and hunter. Leroux led Captain Lorenzo Sitgreaves and the U.S. Army Corps of Engineers. Samuel Woodhouse was also a part of this expedition, and was the first naturalist to visit the area (Ashworth 1997). They camped near Big Leroux Spring, on the southwest side of the Peaks, and studied the plants and animals there (Olberding 2002).

In 1853, Leroux led Lt. Amiel Weeks Whipple and his railroad surveyors on an expedition in search of a route for a railroad through the region to California; they also camped at Leroux Springs.

In 1857, Lt. Edward F. Beale surveyed the region for a wagon road to California. He constructed a road which became known as the Beale Wagon Road. He described grama grass grasslands at the base of the Peaks, and lots of timber and water from many springs.

He said, "it is the most beautiful region I ever remember to have seen in any portion of the world. A vast forest of gigantic pine, intersected frequently by extensive glades, sprinkled all over with mountain meadows and wide savannahs, filled with the richest grasses, was travelled by our party for many successive days" (Ashworth 1997). In 1859, construction of a wagon road was begun.

In 1869, General William Palmer also made a survey through this area to find a route for extending a railway from Kansas to the Pacific (Ashworth 1997). As his party returned through the area heading east, they found abundant water: the lowlands in Fort Valley were 3 ft (1 m) deep in water, the Rio de Flag was full of water and water was seen pouring down the sides of the Peaks (Fig. 2.6). In the 1860s and 1870s, even more explorers, surveyors and prospectors came through the area.

1889: Enter C. Hart Merriam, Life Zones, and the first ecological study of the Peaks

Biologist C. Hart Merriam stepped off the train in Flagstaff on 26 July 1889, and when he looked up at the Peaks, which he was about to study, he saw a mountain whose forests were still strongly structured by frequent fire. Over the course of several months, Merriam described the Peaks' different life zones, or bands of different forest species that replaced one another with elevation. Merriam was ostensibly after a larger, practical goal, holding that a knowledge of life zone boundaries would provide a key to agricultural possibilities in all parts of the United States, including many that were still being settled. Merriam worked for the Department of Agriculture, directing his own agency, which would ultimately become the Bureau of Biological Survey in 1906.

After outfitting the trip, Merriam, his wife, Elizabeth, and assistants Vernon Bailey and F.H. Knowlton headed to Little Spring on the west side of the mountain, where they established a base camp at about 8,200 feet (2,500 m). They made a second base camp just below treeline on Humphreys Peak, and visited its rocky summit several times.

I made a trek up to Little Spring in August. It was late afternoon, very green and threatening rain (Fig. 3.18). The path to the spring is surrounded by forest and small meadows. Elderberries and Canadian white violets were blooming, and water-loving species like cutleaf coneflower were getting ready to. It is a very lush place; the spring was running and there was a large pool of water at its base. How lovely it must have been in the late 1800s, when the forests were more open, with large expanses of grasslands. Fire suppression had only recently started, so the forests were still very healthy.

Merriam made trips between the lower elevation Painted Desert and Grand Canyon and the top of the Peaks, giving him perspective to consider the organization of the region's vegetation along an elevational gradient of more than 7,000 feet (2130 m).

Merriam assigned life zones from the desert to the tundra on the Peaks. In order of increasing elevation from the Painted Desert and Grand Canyon, he described a desert zone below 6,000 ft (1,800 m), the pinyon zone to 7,000 ft (2,100 m), the pine zone to about 8,200 ft (2,500 m), the balsam fir zone to 9,200 ft (2,800 m), the Engelmann spruce zone to 11,500 ft (3,500 m), a narrow treeline or krummholz zone at the top of that, and finally the rocky tundra zone above treeline. Some of his categorizations of the zones and their elevations have changed with more recent studies (see below; Table 3.2). Merriam

3.18. Little Spring, on the west side of the Peaks, was base camp for the Merriam expedition in 1889.

initially assigned names like Hudsonian (for spruce) and Canadian (for balsam fir) to describe the spruce and fir zones, reflecting his thinking that life zones were organized similarly across elevational and latitudinal gradients in North America. He felt that the different plant zones were organized largely by temperature. He described the zones as made up of species that originated in the northern boreal region and in more southern latitudes of North America. The topic of distribution of plant communities was of great interest among scientists at this time; by 1889, more than 50 research papers on this subject had been published (Daubenmire 1938).

Merriam selected 'San Francisco Mountain' for study because of its southern position, isolation, great height and proximity to regional deserts. He gained insight into the organization of ecological communities from the work of Alexander von Humboldt, a brilliant biogeographer, who influenced Charles Darwin as well. Von Humboldt travelled in the Americas between 1799 and 1804, and came to understand the influences of elevation, latitude, soil type and other parameters on plant communities in mountain ranges in the tropics. He regarded plants as the most important signature of the environment (Romanowski & Jackson 2009). The influence of von Humboldt on both Darwin and on Merriam is pretty characteristic of the way scientific discoveries work: few studies or great discoveries occur in a vacuum, but rather are influenced by previous and ongoing efforts, by discussions and debates and ideas.

Merriam was probably the first to recognize the Peaks as a sky island, recognizing the remarkable distribution patterns of tundra plants that occurred above treeline on the Peaks, with many of them also occurring in the Arctic, and even on other continents (Merriam 1890). He pointed out that tundra plants of the Peaks were separated by more than 250 mi (400 km) from related populations in the Rockies, and by thousands of miles in the Arctic, and pondered just how the tundra plants got to the mountain. He knew that they must have migrated to the Peaks during the Pleistocene, and understood that the onset of the glacial phases was slow enough that species were able to extend their ranges southward, avoiding extinction.

In describing the elevations of the zones on the Peaks, Merriam noted that aspect, or the direction in which forests were facing, affected their elevational range: on average, plants occurred as much as 900 feet higher on warmer southwest-facing slopes versus northeast-facing slopes. This pattern is not obvious on the vegetation map constructed for the Peaks (Fig. 4.1, Plate 4). This is a study that should be repeated.

Merriam proposed that the minimum temperature during the growing season, when plants and animals reproduced, determined their distributions on the mountain. As earlier biologists suggested that for birds the temperatures of the breeding season were more critical than at other times of the year, Merriam asked, "if this is true for birds, why is it not also true for other forms of animal life and of plants as well?" (Merriam 1890). Winter temperatures were considered of little importance, as "freezing does not hurt most plants when not in a state of reproductive activity." He suggested that the northern distribution of species was limited by the sum of mean daily temperatures during the growing season, and that the southward distribution of species was regulated by the mean temperatures of the warmest six weeks. Rather meager temperature records were compiled for the project, from as far away as Mount Washington (New Hampshire), as local temperature data were unavailable.

He asserted that temperature rather than moisture limited species' distributions. His reasoning was that while plants and animals might select wetter or drier habitats within a particular zone, they were restricted ultimately to that zone as determined by temperature: temperature "fixes the limits beyond which species cannot pass" (Merriam 1890). No quantitative precipitation data were presented to test this assertion, and the elevational variation in levels of precipitation was ignored (Daubenmire 1938).

Initially, Merriam's work was accepted by many biologists as a great step forward in using plants and animals as the basis for provinces or life zones. In the next few decades, however, plant and animal studies refuted Merriam's suggested temperature laws. While there was support for his description of zones on the San Francisco Peaks, his attempts to extrapolate that relationship across latitudinal and longitudinal gradients elsewhere in North America were not supported. Additionally, neither field nor laboratory tests were used to substantiate his temperature hypothesis. His correlation of a lower threshold temperature of 42 °F (5.5 °C) for growth in all plants and animals was invalidated, in light of very different thresholds for a variety of species (Daubenmire 1938). Furthermore, Merriam admitted that through misunderstanding of Weather Bureau data, 32 °F (0 °C) had actually been used as the threshold value in the summations, rather than 42 °F (5.5 °C). The agreement of zones with such faulty temperature measures was proof of the invalidity of his summation 'law' (see Daubenmire 1938). The error in Merriam's work came about through trying to explain distributions of plants and animals with a single environmental factor.

Some criticism of Merriam's work was valid and some of it was probably just inevitable suspicion of new ideas, which even Darwin couldn't avoid when he suggested natural selection as a driving force in shaping life on Earth. Nonetheless, even Merriam's critics recognized the importance of his work: "His was the first major attempt to use climatic data in interpreting the distribution of the North American biota, and to base distributional provinces on both plants and animals" (Daubenmire 1938). His work inspired considerable research in the field of environmental forces that influence the distribution of plants and animals, including some great studies on the Peaks. Merriam lived large, he named the life zones of the San Francisco Peaks, and generally explored the wild world of this region very passionately. For these things I admire him tremendously. It must have been an amazing trip!

The Zones redefined

Later studies have revised the life zones on the Peaks and elsewhere on the Colorado Plateau, based on a greater understanding of how these communities are organized (Pearson 1920a, b, 1931, Fisher & Fulé 2004, Vankat 2013, Brown 1994, Spence et al. 1995, 2011).

The organization of life zones used in this book is inspired by Spence et al.'s recent classification of plant communities on the Colorado Plateau (2011), and the work of Vankat (2013), Pearson (1931) and Fisher and Fulé (2004), among others. It is as follows: pinyon-juniper zone; ponderosa pine zone; mixed conifer zone; aspen forests within mixed conifer and spruce-fir zones; spruce-fir or subalpine zone; and the alpine or tundra zone (Table 3.2). Some classifications combine ponderosa stands and mixed conifer stands into one montane zone. I keep them separate here because they are so different on the Peaks. The mixed conifer zone has been referred to as the transition zone, the balsam fir zone, the Douglas-fir zone, and the montane zone, among other things. I am sticking with 'mixed conifer' because that best conveys how diverse this zone is (see Mixed Conifer zone below). Fisher and Fulé (2004) and Vankat (2013) considered aspen separately from the other zones because on the Peaks, and elsewhere, it has an enormous range on the mountains, extending widely within both mixed conifer and spruce zones. So, this book also considers aspen separately, not as a zone, but just recognized and discussed separately because it is so fabulous. While Merriam assigned separate zones for spruce-fir forests and for timberline spruce forests, timberline forests are counted as part of the spruce-fir zone here. Bristlecone pine is discussed separately, though technically it occurs with Engelmann spruce at the top of the spruce-fir zone, at treeline, and occasionally above. The pinyon-juniper zone is included sometimes for comparison; it is not currently represented on the Peaks, though of course that will probably change.

The vegetation map for the Peaks (Fig. 4.1), and the patterns that it reveals are presented in Section 4.

1908: Enter Gustav Pearson and his test of the Peaks' Life Zones and Temperature

Gustav Pearson arrived in Flagstaff in 1908 to lead the first U.S. Forest Service research station in the United States. They couldn't have picked a better scientist for the job. Pearson was incredibly well-trained, and utterly driven in his desire to understand how the Peaks' forests worked ecologically.

Pearson must have been inspired by Merriam's recent work on the Peaks, since he set about conducting studies on the Peaks' physical features and its plant communities along an elevational gradient between the ponderosa pine zone and the tundra, and even transplanted many of the forest trees to different elevations on the mountain to experimentally consider their ability to live in other zones. Over the course of several years, he measured temperature, precipitation, evaporation and soil conditions in each of the plant zones. He transplanted ponderosa pine, Douglas-fir, bristlecone pine and Engelmann spruce seedlings among the different zones.

Pearson (1931) showed that temperature decreases and moisture increases as you

Table 3.2. Life Zones on the San Francisco Peaks, past and present (Merriam 1890, Brown 1994, Pearson 1920a,b, 1931, Fisher & Fulé 2004, Spence et al. 2011, Vankat 2011). Area data courtesy of John Vogel, DOI geographer, and 2005 USGS GAP data. Plant species numbers from Appendix I. Species counts per zone do not include mosses and liverworts. Metric in parentheses.

Zone		**Species**	**Elevation feet (m)**	**Area acre (ha)**	**# Plant Species**	**# spp./ acre (ha)**
Current	***Merriam***					
Pinyon-juniper woodland		pinyon-juniper	5,000–6,000 ft. (1,524–1,830 m)	na		
Ponderosa pine	Transition	ponderosa pine, bristlecone pine, limber pine, aspen	6,000–8,000 ft (1,830–2,440)	26,360 ac (10,670)	361	0.014 (0.03)
Mixed conifer	Canadian	aspen, Douglas-fir, limber pine, Southwest white pine, blue spruce, white fir, ponderosa pine	8,500–9,500 ft (2,440–3,505)	14,400 ac (5,825 ha)	342	0.024 (0.06)
Spruce-fir	Hudsonian	Englemann spruce, corkbark fir, aspen, bristlecone pine	9,500–11,500 ft (2,990–3, 505)	10,600 ac (4,290 ha)	231	0.022 (0.054)
Tundra/Alpine	Arctic	nearly 100 plant species	11,500–12,633 ft (3,505–3,850)	1,200 ac (485 ha)	96	0.080 (0.20)

go up the mountain. Evaporation rates are greatest at lower elevations and least in the upper forests (Table 3.3). Intense drought occurs in the ponderosa pine zone in May and June, but not in the upper forests, where soils are typically still moist when monsoon storms begin in early July. The daily range of temperature is greater in ponderosa pine forests than the upper forests (Table 3.3). The ponderosa pine zone—at the bottom—while having the highest daytime temperatures, actually undergoes an inversion of temperature at night, as cold air descends down the mountain, making it colder than higher elevation forests at night. For this reason, Douglas-fir fared poorly when transplanted into the ponderosa pine zone. Shady versus open microsites sites affect these measures in all of the zones, extending a species' range up or down in elevation.

Temperature and water

Pearson concluded, after studying transplanted trees from 1918 to 1926, that the ranges of some of the forest trees were limited by temperature *and* water: many species were limited by cold temperatures at upper elevations and by limited moisture at lower elevations (Pearson 1920b, 1931). *Ponderosa pine* appeared strongly limited by cold winter temperatures in the highest forests, and at the other end, able to persist even in lower pinyon juniper woodlands—with enough water. *Douglas-fir* of the mixed conifer zone was readily damaged by early summer frosts and by winter frosts, and

Table 3.3. Pearson (1931): San Francisco Peaks moisture, temperature and evaporation by life zone. Metric in parentheses.

Zone	Annual Precip. In/(mm)	Air Temp. Mean °F/(°C) Ann.	Air Temp. Mean °F/(°C) June–Sept.	Summer % Soil moisture in first foot	Summer Evap. In/(mm)	# Frost free days (1918)
Pinyon-Juniper	12–20 (304–508)	53 (11)	69 (20)	2.6	30–40 762–1.016	140
Ponderosa pine	18–25 (457–635)	46 (7.8)	61 (16)	3.8	18–26 (460–660)	116
Mixed conifer	22–34 (558–863)	41 (5)	57 (10)	13.7	12–16 (254–330)	140
Alpine tundra	40* (1,016)	—	48 (9)	—	—	114

* 1919 only

fared poorly when planted in the spruce-fir zone. It is noteworthy that the mixed conifer zone in the mountain has a frost-free period similar to that of the lower elevation pinyon–juniper woodlands and yet gets nearly as much moisture as the spruce-fir zone above (Table 3.3). For these reasons, Tom Kolb, NAU forest ecologist, regards the mixed conifer zone as the 'sweet spot' on this mountain. Pearson found that Engelmann spruce and bristlecone pine were limited by temperature and moisture at upper elevations on the Peaks, and by limited moisture at lower elevations.

Pearson's study showed that colder temperatures in winter *and* summer could limit the upper distribution of some species on the Peaks, as with ponderosa pine's absence from the highest zones in the mountain, and Douglas-fir's failing in the ponderosa pine zone. Merriam insisted that it was colder temperatures during the summer rather than winter that limited species (see above). Indeed, transplanted ponderosa pines were not able to grow much during the shorter growing season at higher elevations—shorter by at least a month—and failed there after several years. And yet, Pearson also found that winter temperatures, along with winds and solar insolation, were very damaging to some of the lower elevation trees when transplanted to higher elevations. Greater shade in the spruce-fir zone also limits the presence of shade-intolerant trees such as ponderosa pine there.

Pearson found that water availability, rather than temperature, limited the lower elevations at which many of the species could occur. He noted that a number of trees from the Peaks' upper zones, including Douglas-fir, limber pine and spruce, were growing quite successfully at 7,000 ft (2,130 m) in nearby Flagstaff, and in wet canyons carved into the Mogollon Rim down to nearly 5,000 feet (1,500 m)! The deserts might be another story. So plants can move down out of the mountains in warm times if there is water. This story is still unfolding.

There is talk of repeating Pearson's experiments, and efforts are underway to locate his sites and any remaining transplanted trees!

Fire-Driven Mountain Ecology

While Merriam and Pearson studied the effects of water and temperature on the Peaks' plant communities, there were other enormous forces at work on the mountain, which, it turns out, may be even more important than climate in shaping the Peaks' life zones. Through some wonderful studies, recent research has shown that these forces profoundly affect the mountain's ecology in very important ways. Of course, the first force is fire.

Now seems like a good time to mention that the Southwest has the highest concentration of lightning-caused fires in the continental United States (Barrows 1978, Pyne 2001, Vankat 2013). Arizona leads with 52 lightning ignitions per 1,000 miles^2 per year (52/2,600 km^2/year; Kay 2007, Vankat 2013). Ignition rates are even greater in many of Arizona's mountains (e.g., 5,000/1,000 mi^2, 5,000/2,600 km^2 per year; Swetnam & Baisan 2003).

Fire plays a huge ecological role in the Peaks' forests. During monsoon storms, numerous lightning strikes occur all around the mountain (Fig. 3.19). Most fires start during the onset of monsoon storms each June into July—80% of all fires, to be exact, occur during this time (Vankat 2013). This follows on the heels of traditional May-June drought, when temperatures typically reach their peak for the year. Because of this heat, highly energetic storms occur near the Peaks each summer. The season's first thunderstorms are often accompanied by only virga or sparse rain and 'dry lightning', which actually causes the most fires (Allen 2002). Lower, drier ponderosa pine forests sustain more than 80% of these fires on the Peaks (Barrows 1978). As a longtime resident of Flagstaff, at the base of the Peaks, I can attest to great anxiety during this time of year, wondering if this will be the year when the big fire occurs. Mixed conifer zones don't get as many lightning-caused fires as the ponderosa pine zone below. They account for 12% of fires, compared with the 1% that the spruce-fir zones get (Vankat 2013).

Tree rings, fire scars and the history of fire on the Peaks

Several tree ring studies on the Peaks provide evidence of frequent fires in its forests as far back as the 1500s (Fig. 3.20; Heinlein et al. 2005, Cocke et al. 2005, Margolis et al. 2011). Tree ring patterns or dendrochronology provide ages, sometimes back more than a thousand years, while fire scars can tell how frequent and widespread fires are in a forest. Together they describe how forests were structured, how open or dense they were and how they were changed by fire. Information from regional lakes established a history of local forests dating to the late Pleistocene and early Holocene, with some indication of fire through the last 50,000 years, but it's the tree ring studies that detected the many, many fires that the forests on the Peaks have experienced. They also tell the story of huge changes in the forests starting in the late 1800s.

Fire and forests on the Peaks

Heinlein et al. (2005) studied tree rings in mixed conifer stands on the east and west sides of the Peaks, and found that for hundreds of years, up until the late 1800s,

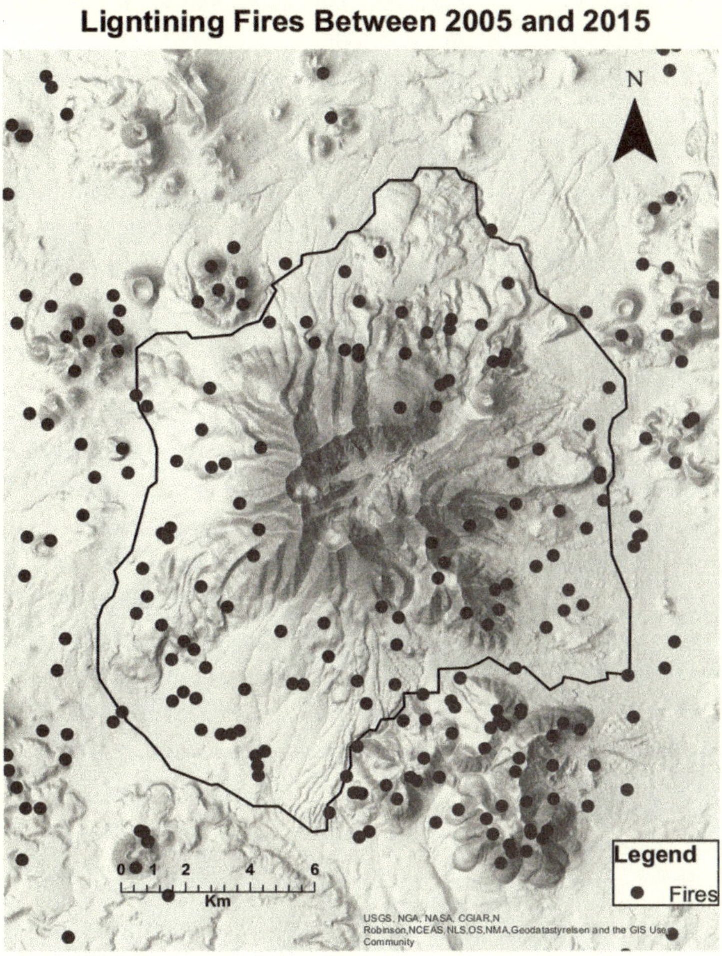

3.19. Lightning-caused fires around the Peaks between 2005 and 2015. Courtesy of Ian Heynssens and the USGS.

they were open forests, with densities of trees ranging from 43 to 60 trees per 2.5 acres (1 hectare). The openness was due in large part to frequent ground fires, started by lightning and carried by grasses, which eliminated most conifer seedlings. They found that between 1500 and the late 1800s there were more than 250 fires in the mixed conifer stands they studied, averaging one fire every 5 years (Fig. 3.20). Nearly half were large fires, and most occurred in summer. It's not surprising that these stands were dominated by ponderosa pine, which is more tolerant of fire than Douglas-fir, limber pine, white fir, and aspen. Also, shade-intolerant ponderosa pine found more light in these open forests.

Another tree ring study considered the response of spruce-fir forests to a stand-replacing fire on the Peaks (Margolis et al. 2011). According to this study's tree ring dates, the massive stand-replacing fire in the Inner Basin burn featured in a 1910 photograph probably occurred in 1879 (Fig. 3.21). In the photograph, huge expanses of spruce-fir forest and aspen stands were burned to the ground, leaving only a few fins of unscathed conifers alive within the burn area. Judging by the remaining burned stems and their great densities, it must have been a pretty dense conifer stand, with aspens marginalized. The interesting thing about comparing this photograph with the current forest, is that both aspens *and* conifers recovered remarkably quickly after this very destructive fire—regardless of when exactly it burned. I wonder if seed rain from conifers that survived in the basin, especially those that survived on some of the upper slopes, expedited recolonization.

This burn resulted in a major establishment event for aspens. Margolis et al. (2011) showed that 63% of all aspen trunks sampled in the Inner Basin site were established within 5 years after the 1879 fire. This is a wetter, cooler area (Fig. 3.4). These stands have thrived on moraines associated with the Core Ridge glacier, and an arm of it which extended down from Doyle Canyon (Fig. 3.1). There were still few conifers in some of these aspen stands as of 2016, some 137 years later, which shows how the clonal habit of aspen, along with its other wonderful traits, makes it a winner in a high elevation, fire-prone area. These aspen stands have gotten to thrive in the sunlight, uninterruptedly, for more than one hundred years, and will continue to do so until conifers make their way into their stands via wind-blown seed, grow slowly in their shade for many years, and finally overtop them.

As mentioned, the conifers have also recovered pretty fast, filling in a lot of the Inner Basin burn site within a little more than 100 years. While perusing the local USFS photo archives for this book, I discovered that the Forest Service actually planted Engelmann spruce seedlings in the Inner Basin in 1911. So, perhaps that explains their rapid re-establishment in some areas there.

In addition to keeping forests open and healthy, frequent fires also kept the more fire intolerant trees, like spruce and fir and Douglas-fir, confined to the Peaks' upper forests, as most fires occurred in the lower forests. The general thinking is that fires were most common in the lower ponderosa pine stands, with rates as frequent as every several years. Most of these were fast-moving ground fires that cleared debris and seedlings, with little damage to existing pines. In mixed conifer stands, fires were thought to be of mixed severity, with some fast ground fires occurring, and also larger, more destructive

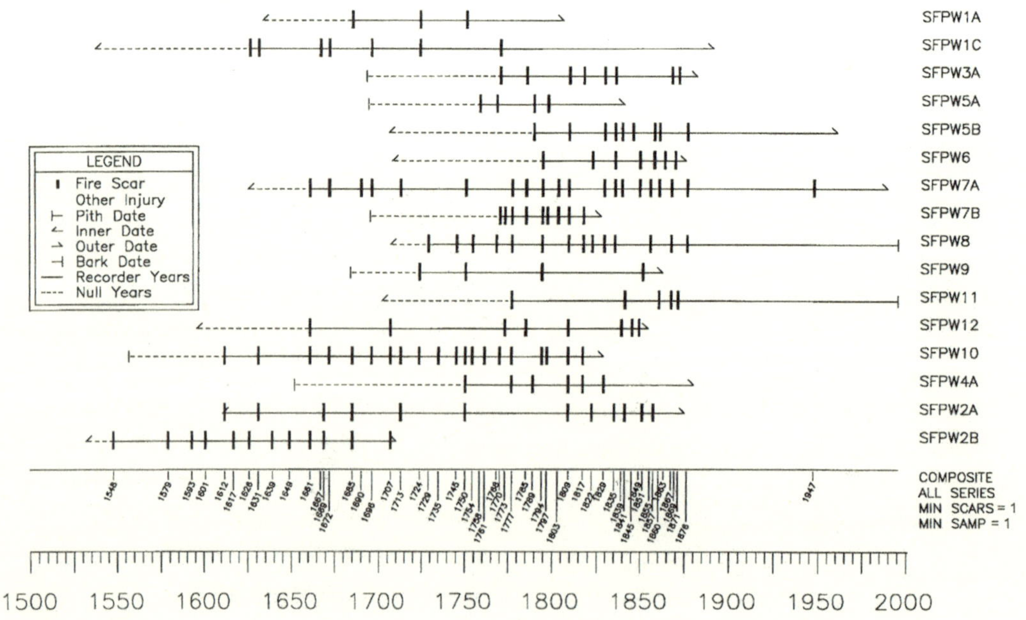

3.20. Fire history chart for the west side of the Peaks. Note the abrupt cessation of fires, indicated by vertical hatchmarks in the late 1880s. Courtesy of Thomas Heinlein and the International Journal of Wildland Fire.

3.21. Photo of a stand-replacing fire in the Inner Basin. The upper photo was taken in 1910, by A.G. Varela, USFS, and may represent the aftermath of a stand-replacing fire in 1879. The lower photo was also taken from Agassiz Peak in 2016.

fires. Fires in the spruce-fir zone were and are thought to be less frequent, but also the most severe, with stand-replacing fires, such as the Inner Basin fire. They all created open and healthy forests.

Grazing And Fire Suppression

Fire would be nearly eliminated in the late 1800s; only two fires were reported in these stands between 1892 and the mid 1900s (Fig. 3.20).

Loss of natural fire in a strongly fire-prone region has had profound ecological consequences. Resulting denser forests with greater ground debris combined with warming summers in the Southwest have led to some large fires, unprecedented in scope, intensity and destruction (see following page, Table 3.4).

The short version is that overgrazing suppressed fires which led to denser forests and a buildup of ground debris which has led to catastrophic fires, as have been seen during the 21st century—who knew? To be perfectly clear, the removal of grasses by overgrazing eliminated the fine fuels that carried frequent, cooler ground fires. This loss contributed most significantly to the abrupt end of fires on the Peaks, and elsewhere. Without these fine fuels, lightning strikes were less likely to expand into fast-moving ground fires that kept things open and healthy. It was overgrazing that led to most fire suppression, more than suppression efforts by people. This kind of came out of nowhere, a consequence that no one was expecting.

So, the introduction of massive herds of livestock into the fire-driven mountain ecology of the Peaks starting in the late 1800s had enormous, long-term impacts on its forests. People began to settle in nearby Flagstaff in 1875, primarily to raise livestock, because of the area's reputation for unlimited grasses and few fences (Olberding 2002). The

Table 3.4. Recent fires on the San Francisco Peaks. Courtesy of J. Walther and Coconino National Forest.

Site	Year	Acres	Cause
Inner Basin	1879	>500	lightning
Fort Valley	1948	1,920	
Hostetor	1950	50	human
Hostetor	1968	160	
Kelly	1971	3,550	human
Wild Bill	1973	7,140	human
Schultzer	1973	15	human
Burnt	1973	7,150	human
Dry	1974	10	lightning
Schultz	1977	15	human
Brookbank	1977	12	lightning
Elden	1977	4,594	human
Weatherford Trail	1980	70	lightning
Hochderffer	1980	50	lightning
Aspen	1981	12	lightning
Bear Jaw	1995	900	human
Bismarck	1996	10	lightning
Horseshoe	1,996	8,650	human
Hochderffer	1996	16,400	lightning
Doyle	1996	50	lightning
Elden	2000	10	lightning
Leroux	2001	1,200	human
89 Mesa	2010	523	human
Schultz	2010	15,075	human
Hart	2015	15	lightning

arrival of the railroad in 1883 really helped launch emerging grazing and logging efforts in the area into overdrive (Ashworth 1997, Cline 1982, Olberding 2002).

Historian Platt Cline (1982) suggested that by 1891 there may have been 200,000 sheep grazing on the Peaks alone, while surrounding areas experienced even greater numbers. He estimates that up to one and a half million head of livestock grazed in northern and central Arizona by the early 1890s. Often livestock were left in place for long periods of time, removing whole sections of the lush grasses known to the region, devouring them and uprooting them and disturbing the soil (Fig. 3.22, Cline 1982, Leiberg 1904, Ashworth 1997, Pearson 1931).

The presence of many carvings on aspens by Basque shepherds in mixed conifer and spruce-fir forests indicates that most forests in the mountain were grazed, and particularly by sheep, during the 1800s through the mid-1900s (Fig. 3.22). Coconino National Forest has cataloged hundreds of Basque carvings on aspen trunks on the Peaks (Fig. 3.23). They are found throughout the mountain's aspen groves and spilling down towards prairies such as Hart Prairie, on the west, east and north flanks and through much of the Inner Basin, with recorded dates ranging from 1882 to the early 1980s. Not surprisingly, sites with springs were particularly popular, including

Leroux Spring and Lockett Meadow Spring. A friend recounted being up at Hart Prairie in the 1980s and seeing huge flocks of sheep working their way up from the prairie into the aspen stands above.

So, frequent fires ended on the Peaks in the late 1800s (Fig. 3.20), and the ecological ramifications have been profound for open, healthy mountain forests that had been strongly organized by fire for as long as there have been forests here. The same pattern of abrupt fire disruption occurred in many southwestern forests during this time (Swetnam & Baisan 2003, Fulé et al. 2003a, 2003b). In northern Arizona and New Mexico, the last fires were reported at Camp Navajo in 1882 (Fulé et al. 1997), on the south rim of Grand Canyon in 1887 (Fulé et al. 2003a), and between 1881 and 1892 in northwestern New Mexico (Grissino-Mayer & Swetnam 2000). The fire disruption date of 1876 on the west side of the Peaks coincided with the introduction of large numbers of livestock into the Fort Valley and Chimney Spring area (Fig. 2.6; Heinlein et al. 2005). Other human land use, including trails, roads and railroad lines, disrupted fuel continuity and also contributed to the cessation of natural fire regimes.

JOHN LEIBERG AND THE PEAKS IN 1901

In 1901, in between the studies by Merriam and Pearson, John Leiberg of the United States Geological Survey arrived to assess the condition of the the Peaks and surrounding lands, at a time when overgrazing, overlogging and environmental destruction were occurring through much of the western United States. Based on recommendations of Bernard Fernow, one of the early heads of the emerging U.S. Forest Service, these areas were designated as Forest Reserve lands by President McKinley in 1898; and in 1902, President Theodore Roosevelt consolidated all of the tracts into the San

3.22. Sheep grazing near the Peaks. Courtesy of Ecological Restoration Institute, NAU.

3.23. 'Dendroglyph,' a Basque shepherd's etching on aspen, while grazing sheep on the Peaks. Courtesy of USFS.

Francisco Mountains Forest Reserve (Leiberg 1904). Thus, these public lands came to be managed by the US Forest Service (Olberding 2002).

Leiberg's work reflects the clearest understanding and expression of conditions on the Peaks during this profoundly dynamic time. Based on a two-year study, Leiberg gave the reserve lands a very poor grade, describing the destruction of their forests and grasslands due to grazing, fire and logging (Leiberg 1904). He described many areas around the Peaks in which ponderosa pine had been logged to 'exhaustion,' with up to 100% of trees removed in places. This contributed to a lack of recruitment of ponderosa pines in the following years. Leiberg noted the recent decrease in fires in the reserve, despite evidence of many previous fires. He estimated that approximately 18,000 acres (7,300 ha) of forest had burned on the western, southern and northern slopes of the San Francisco Peaks.

Leidberg observed that the most extensive and destructive fires had occurred in mixed conifer and spruce-fir forests on the higher slopes of the Peaks and Kendrick Mountain and Sitgreaves Mountain, rather than in ponderosa pine forests. He regarded ponderosa pine as the most fire tolerant conifer in the 'Pacific United States.' So, while frequent ground fires were started by lightning in the vast stands of ponderosa pine surrounding the mountain, these fires were most often cool ground fires, while the higher elevation forests carried larger, destructive fires, at less frequent intervals.

Leiberg aptly described the nature of the earlier, more natural fires in ponderosa pine stands: "Owing to the heavy grass growth which prevails on all areas not sheeped off, surface fires develop considerable heat and flame and death is certain to all seedling growth..." These frequent fires kept ponderosa pine stands open, healthy and diverse. He recognized that most fires were initiated by lightning strikes, with up to 50% of mature ponderosa pines being seriously damaged or killed by them in some sections.

Sheep and cattle destroyed sprouting aspens and pine seedlings, along with the diversity of native grasses that occurred in the region. Leiberg described damage done by "bands of sheep being herded back and forth over much or all of the reserve" (Leiberg 1904). Their damage was as much related to extensive trampling as to grazing. "They pulverized the soil to a dust-like consistency, which was washed away in rain storms..."

As recently as the early 1970s, herds of up to 2,000 sheep grazed in Lockett Meadow in the Inner Basin (Haldemen et al. 1973).

Changing Forests On The Mountain

These changes were afoot on the Peaks, though Merriam apparently didn't see or didn't regard them as germane to his views on the organization of life zones on the Peaks. Ultimately, overgrazing and fire suppression have challenged Merriam's assertion that the life zones on the mountain were organized solely by temperature. Even Pearson stayed highly focused on the primary questions about elevation in his study, often fencing his transplant experiments to protect them from cattle and sheep (Pearson 1931).

Zones reconfigured by a lack of fire

Studies by Northern Arizona University's Ecological Restoration Institute (ERI) found that cessation of fire enabled the mountain conifers, including bristlecone pine,

Engelmann spruce and various mixed conifer species, to expand their ranges considerably downslope—down elevation—below their zones into mixed conifer and even ponderosa pine forests (Cocke et al. 2005, Heinlein et al. 2005). Essentially, the absence of fire for the last hundred years has blurred the more discrete zones that Merriam saw, indicating that fire even more than temperature had been limiting the downward distribution of these mountain species.

According to Alison Cocke et al. (2005), there has been a shift in species composition towards greater dominance of mesic species at lower elevations in the absence of fire since 1876. In the 1876 forest, spruce, fir, bristlecone pine and limber pine were absent from the ponderosa pine zone, while they all occur there today. A problematic aspect of fire-intolerant trees in lower zones is that they have made these fire-prone habitats even more susceptible to fire, including larger, catastrophic fires as experienced today (see below). Many of these species lack fire-tolerant features, such as thick bark. By 2000, bristlecone pine had descended some 500 ft (150 m) in elevation, Engelmann spruce descended 1,300 ft (400 m), Douglas-fir 180 ft (55 m), and limber pine more than 80 ft (24 m), when compared with their lowest elevations in 1876. Today bristlecone pine occurs more than 3,000 feet (914 m) below its upper distribution and can be seen thriving in the open slopes of Hart Prairie—along with ponderosa pine, on the west flank of the Peaks. It has also become more prominent in the mixed conifer zone.

Too many trees

This movement of fire intolerant conifers down into lower, warmer zones was facilitated by a great wave of establishment of forest trees throughout the mountain, also a result of fire suppression; and densities continued to increase significantly during the next 124 years, as measured in 2000 (Cocke et al. 2005). Densities increased most in ponderosa pine stands with a 1073% increase (see ponderosa pine section below), and least in bristlecone pine (92%)! Heinlein et al. (2005) found that densities in mixed conifer study sites increased from 43 to 60 trees to between 928 to over 1,700 trees per 2.5 acres (~ 1 hectare) today! Changes in tree density and basal area were least in the spruce-fir and bristlecone pine zones. Often these forests become extremely dense and dark, and are referred to as 'doghair thickets': they lack diversity and are unhealthy (Fig. 3.24).

From cool ground fires to catastrophic stand-replacing fires

Cessation of fires in the last 100 years has actually increased the potential for more catastrophic, stand-replacing fires in the Southwest. Historically, most fires through ponderosa pine forests were thought to be fast moving, cool fires sometimes with residence times of less than a minute. They cleared the ground of organic debris and pine seedlings, and left forests open and healthy. The grasses that provided the fine fuels that carried such fires were eliminated by overgrazing. Their removal also eliminated competition for establishing pine seedlings, leading to a massive increase in trees in the last 100 years (see above) and an enormous buildup of fuel from needles and branches and cones on the ground. The high densities of trees and increasing ground fuel are contributing to a trend of larger fires and more destructive fires—stand replacing fires (Table 3.4, Fig. 3.20; Heinlein et al. 2005). These may become the norm in ponderosa pine forests,

3.24. A doghair thicket of ponderosa pine established with its 1919 population explosion.

particularly as the less fire tolerant trees move down into that zone. With global warming and drought predicted for the region, this is a really troubling trend.

The latest and the largest recorded fire on the Peaks was the Schultz Fire of June, 2010; it burned 15,000 acres (6,070 ha) on the eastern side of the Peaks (Fig. 3.25, Table 3.4). Nearly 60% of the area burned was burned in one day, due to relentless winds and a dense dry forest. The fire burned across moderate to very steep slopes of ponderosa pine and mixed conifer forests. Most of the slopes experienced high severity fire: nearly all trees were killed in this stand-replacing fire. This summer fire was soon suppressed by unprecedented monsoon storms. Storms dumping up to an inch of rain (25 mm) per 15 minutes occurred several times, and led to enormous erosion and flooding, one mortality, and considerable property damage (see Water section below). I know a family who lived in the path of the flooding who literally had to drop everything and run as a wall of mud and water descended on their home.

3.25. The extent of the 2010 Schultz Fire on the east side of the Peaks. Top: Peaks on fire at night; Middle: The lighter terrain reflects a near total burn on the east side of the Peaks. Bottom: Burned drainage on Schultz Peak slopes.

(Following page) Top: Bitter chokecherry holding down post-fire slopes; Middle: Resprouting aspens overtop hikers 7 years after the Schultz Fire; Bottom: Ponderosa pine seedling establishment. Fire image and burned drainage image courtesy of Coconino County.

— SECTION 4 —

Now

Who Lives On This Sky Island Today...

The Peaks' four plant zones cover 56,732 acres (23,000 ha, Fig. 4.1, Plate 4, Table 3.2). Today, 567 taxa of plants are known from the Peaks (Appendix I). There are also more than 100 species of fungi, 8 species of lichens, 161 species of birds, and 40 species of mammals that live on the mountain (Appendices II-IV). More than 30 species of mosses and 2 liverworts were added to the plant list; however, their distributions are not currently understood on the mountain, and so they are not considered in summaries of species per zone, in this edition. The insect story is coming together, with 352 bees, 96 species of flies and 70 moths and butterflies now documented for the mountain.

Three plant species are endemic to the mountain; they occur nowhere else. They are the San Francisco Peaks ragwort, graceful buttercup and bristlecone pine dwarf mistletoe (Appendix I, Fig. 4.16, Plate 38, 39). Perhaps not surprisingly, the endemics all occur at higher elevations, with their populations being so isolated.

The life zones themselves are really interesting. The Peaks' new vegetation map shows the distribution of the different plant zones or communities and provides a more accurate estimate of their areas or sizes. This map was achieved by considering the mountain as the 3-dimensional form that it is. A few highlights of the map include: the area of the ponderosa pine zone is quite arbitrary, really, as its area around the Peaks is but a small part of its enormous western range. The high diversity of plants in the ponderosa pine zone surely reflect its larger range, and make for an interesting comparison with plant diversity in the upper zones (Table 3.2). Moving up the mountain, the ponderosa pine zone and the mixed conifer zone including aspen overlap extensively between about 8,000 and 8,500 ft (2,440 to 2,600 m). Aspen has an enormous presence on this mountain. It occurs on more than 12,000 acres (4,850 ha), as a dense,

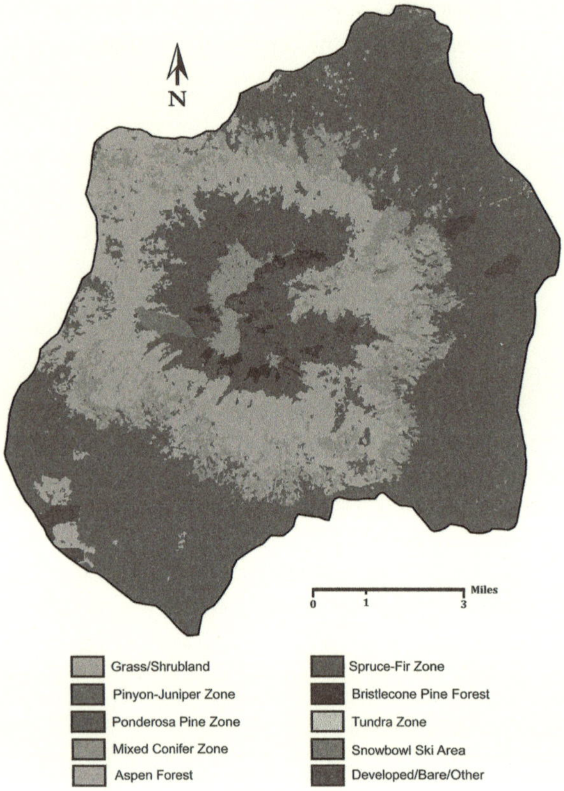

4.1. Vegetation map of the San Francisco Peaks, showing its different life zones and other features. Courtesy of John Vogel, DOI geographer & 2005 USGS GAP data. See Color Plate 4.

nearly continuous band all the way around the mountain, especially between about 9,000 to 10,000 ft (2,740 to 3,050 m, Fig. 4.1, and back cover image). This dense band of this wonderful clonal tree is intriguing. What is going on in that dense hub? Is that where most lightning strikes occur? Or is it perhaps the environment that especially suits aspen? Farther upslope, spruce-fir occurs in larger stands starting at over 10,000 ft (3,050 m). And Engelmann spruce and bristlecone pine occur up to treeline and even a little above. By that point they are small and windswept shrubs (Fig. 4.15). Tundra plants are often quite diverse in these treeline stands (Fisher & Fulé 2004, Schaack 1983).

Most plant species on the Peaks occur in the ponderosa pine zone, in part because it is such a huge forest here, extending around nearly the entire rim of the Colorado Plateau (Appendix I, Table 3.2, Fig. 3.10, Plate 3). There are nearly as many species in the much smaller mixed conifer zone, or the sweet spot. The smallest number of species occurs in the tundra—above treeline. This is expected given the very small area of this zone. But get this: when the number of plant species per zone is divided by the size of that zone, there are actually more species per acre in the tundra than there are in ponderosa pine forests—over 4 times as many, actually (Table 3.2)! Wow, did not expect that. All those species up there. Their numbers surely relate to the lack of trees.

C. Hart Merriam suggested that most plants occurred in single or few zones, and the data for the Peaks support that: nearly three quarters of plants on the Peaks occur in one or two plant zones. Most of those occur in the ponderosa pine and mixed conifer zones. Only 5% of the species (27) occur in all the Peaks' zones.

In Arizona, there are 33 plant species that occur *only* on the Peaks (Appendix I). It's especially remarkable that most of them are high elevation species, including lots of sedges, buttercups and bristlecone pine (*Pinus aristata*, Appendix I). Nearly all of these species hail from the Rockies and Utah, probably migrating to the Peaks during the Pleistocene, and have remained here as the climate has warmed. The Peaks is as far south as they got or at least have persisted.

There are 14 species that stay above treeline on the Peaks, while many more drift down into the spruce-fir forests, but no lower. These are the species that really make the Peaks a sky island. They live at the top of the mountain, utterly isolated by hundreds of miles from related populations. These include the northern buttercup, sticky polemonium (Plate 37), a grass, a rush and 5 species of sedges (Plate 34). They all hail from populations in the Rockies and Utah.

Most plant species on the Peaks have ranges that extend through much of the West, making this mountain awash with species that originated to its north and to its south, species that live along the Western Cordillera, which I like to think of as the western mountain plant highway. I hope someone will study this, and really figure out the shape of the Peaks plant community, within the Cordillera. Anyway, the Peaks is a beacon of the mountain world on the Colorado Plateau, surrounded by enormous expanses of cold deserts and low woodlands through northern Arizona, between it and related populations in the Rockies.

This project has helped to shine a light on the remarkable diversity of plants on *and around* the Peaks, through the San Francisco Volcanic Field. The map drawn around the Peaks for the purpose of making the plant list as Peaks-specific as possible helped to reveal some of the very local distributions of species within the volcanic field. Many plant species occur close to the Peaks, but not on the Peaks. More than 120 plant species occur on neighboring Mt. Elden that do not occur on the Peaks (Appendix V). Even black dropseed (*Sporobolus interruptus*), an endemic grass that occurs only along the Mogollon Rim, does not occur on the Peaks. Some are lower elevation species that favor the south slopes of Mt. Elden or the cold desert; many species reach their northern limits there. A number of cinder cones in the SFVF have unique plant associations, even including endemic species on some of the older cones (Christie 2006). Engelmann spruce occurs on the Peaks and Kendrick, but not Bill Williams. But the bottom line is that this region, with its enormous elevational and geological variation, ranging from low deserts to high deserts to treeline habitats, supports a great diversity of plant species. It may be that many of them will continue to find their place in the region as it warms.

Most conifers on the Peaks occur only in western North America, in its sea of rugged mountains (Fig. 1.1). The conifers

4.2. Female cones of most conifers in the Peaks; left to right: pinyon pine, ponderosa pine, limber pine, Douglas-fir, corkbark fir, Engelmann spruce, bristlecone pine. Ponderosa pine and bristlecone pine cones have spines for discouraging seed herbivores.

on the Peaks include common juniper, Utah juniper, ponderosa pine, Douglas-fir, white fir, limber pine, southwestern white pine, blue spruce, Engelmann spruce, corkbark fir and bristlecone pine (Fig. 4.2, Appendix I). The mostly north-south running mountains of the Western Cordillera have permitted offshoots of the northern boreal forests to extend their ranges all the way to Mexico, and likewise, for more temperate and subtropical species like southwestern white pine and limber pine to travel north! Common juniper occurs through the Northern Hemisphere.

In the process of migrating, most of these conifers diverged genetically, shaped by natural selection along the way. Nearly all of them are represented by multiple subspecies, varieties and populations that are just different. Most of this variation is related to gradients in elevation and latitude in the Cordillera, across which these species have migrated back and forth for a long, long time. Populations along such gradients represent a cline.

So, Andy Stevenson, USFS Coconino National Forest silviculturalist, on the Peaks, says that all the trees on the mountain are showing good 'regen' or regeneration these days, thereby replacing themselves (Andy Stevenson, personal communication 2017). This is great news and the bottom line.

Ponderosa Pine Zone

Walking through a ponderosa pine forest can be very different depending on the particular stand you are in, as its forests are so variable. I love the open forests with old growth trees, dating back sometimes more than 500 years (Fig. 4.3). Their openness results in lush understories of flowers and grasses. By contrast, ponderosa pine doghair thickets are unpleasant to be in: they are stands that are too dense, resulting in dark, unhealthy forests that lack diversity; they feel dead to me (Fig. 3.24).

4.3. Ponderosa pine old growth stand, with a grassy understory.

Today, ponderosa pine has an enormous, contiguous distribution in the Southwest alone, covering nearly 19,000 miles^2 (49,000 km^2, Fig. 4.1; Vankat 2013). On the Peaks, this conifer covers more than 26,360 acres (10,660 ha); this vast forest is far larger than the life zones above it (Table 3.2).

The earliest record for ponderosa pine is from fossils in British Columbia dating back to about 50 Ma (Bonnicksen 2000), and it has been present in the West ever since. According to the fossil record, it disappeared from much of the Southwest during the last glacial maximum of the Pleistocene, due to colder conditions or a lack of summer rains. With warming it colonized western mountains quickly, and for at least the last

11,000 years it has thrived in the West—it has made the West its own, becoming one of the most widespread conifers there. Records from Potato Lake—some 40 miles from the Peaks—documented its absence near the Peaks 35,000 years ago, and then its presence here by about 12,000 years ago (see above; Anderson 1993, Anderson et al. 2000).

A ponderosa pine cline

As ponderosa pine's range has migrated northward since the end of the Pleistocene, pronounced differences have developed among its populations along the way from southern Arizona northward to the Rockies (Rehfeldt 1993). This differentiation happened remarkably fast, as ponderosa pine migrated northward at the end of the Pleistocene, within the last 10,000 years. Gerald Rehfeldt, USFS plant geneticist, found that 85% of the variation in ponderosa pine's growth patterns was explained by latitude and elevation, both of which increase between southern Arizona and the Rockies. By comparing growth patterns in different populations of trees in a common garden experiment, he found that ponderosa pine from lower latitudes, including the Bradshaw Mountains in central Arizona, grew more rapidly, being genetically accustomed to more frost-free days and was less frost-tolerant, while populations farther north in the Rockies were slower growing and more tolerant of cold (Rehfeldt 1993). Rehfeldt found that, though differences among populations between Arizona and New Mexico, and Utah and Colorado are great and may warrant "racial" delineation, the transition between races is quite broad, meaning a more gradual relationship gradation between these populations. Populations separated by as little as 700 ft (229 m) in elevation differed genetically, in response to subtle but significant environmental differences such as a change in the number of frost free days. So fascinating that the responses of plants can be so gradually and finely delineated along such gradients. It turns out that adaptation to climate in plants is a compromise between selection for fast growth and for cold tolerance; optimizing both is not possible in plants (Rehfeldt, personal communication, 2016). Such patterns demonstrate how quickly even ancient conifers can adapt to very specific conditions.

There is hope in this—this ability of a rather primitive plant to adapt to the finer details of its local climate within a pretty short time. So as climates warm, these local populations may continue to adapt...brave new world and all that. Of course, ponderosa pine populations along the Pacific coast and those of the interior West diverged a long time ago in response to their vastly different maritime climate (Latta & Mitton 1999).

Ponderosa pine basics

Ponderosa pine, like most conifers, is dependent on the wind to disperse its pollen, and then its seeds. Ponderosa pine is monoecious, meaning one house, as each tree contains female and male cones, with most female cones in the upper canopy and male cones throughout (Fig. 4.2). It is full-on conifer primitive: female cones take two years to form and then become full-sized seed cones by year 3. Compare this with the reproductive cycles of flowering plants which develop seeds from flowers over the course of one growing season or less. Ponderosa pine's winged seeds aren't all that good at dispersing—I often see clusters of seedlings establishing near the parent tree (Fig. 4.4), making its rapid recolonization

4.4. A cluster of ponderosa pine saplings established near the parent tree.

of much of the West seem surprising. Seeds typically disperse in late summer. Ponderosa pine's seeds typically germinate with summer monsoon storms.

In forests on the Peaks, a ponderosa pine more than 700 years old was found at Fort Valley, and occasionally pines approaching 500 years are found on the Peaks lower slopes (Peter Fulé, personal communication, 2017). The oldest known tree has been dated to about 1,000 years old (Peter Fulé, personal communication, 2017).

Ponderosa pine is intolerant of shade, which partly accounts for its low numbers in the denser forests above. It is also one of the more drought tolerant pines, due to characteristics such as sunken stomates or leaf pores that close quickly with aridity, a thick, insulating bark layer, the ability to store water in its central heartwood, and an array of associated mycorrhizal fungi which help its roots to take up water and nutrients.

Not surprisingly, this pine is strongly fire-adapted: thick-barked older trees tolerate low intensity fires, bearing fire scars as records of lightning strikes and fires, though seedlings and saplings are typically eliminated by such fires. Ponderosa pine densities were kept low by frequent fires in and around the Peaks, before fire suppression started in the late 1800s. One of the shortest fire return cycles in ponderosa pines on the Peaks—1.8 years—was measured in the Chimney Springs area at the base of the mountain (Dieterich 1980).

As a result of fire suppression, ponderosa pine numbers are much greater today. A study on the Peaks found that its densities increased by 1073% between 1876 and 2000 (Cocke et al. 2005, see above)! Increased numbers of trees increase competition for water. A study at Fort Valley Experimental Station at the base of the Peaks found that heavy thinning of the dense pine stands improves water relations for remaining trees (Fig. 2.6; Kerhoulas et al. 2013a, b). This study also found that older pines draw on water sources deeper in the ground supplied by winter water, while seedlings and saplings use shallower monsoon season water (Kerhoulas et al. 2013b).

Gustav Pearson and ponderosa pine ecology

While Gustav Pearson conducted a study of the forests at different elevations on the Peaks, he also intensively studied the relationship between ponderosa pine establishment and climate. He set out to find out why ponderosa pine was not replacing itself in the heavily logged region around the Peaks, and how to improve establishment of its seeds. He studied ponderosa pine's recruitment patterns at the Fort Valley Experimental Station at the base of the San Francisco Peaks between 1908 and 1920, and attempted to establish pine seedlings over vast logged tracts all around the mountain (Fig. 2.6). Through a combination of excellent long-term experiments, and massive planting efforts which ultimately failed, he concluded that natural recruitment, rather than planting seedlings, was the only way that ponderosa pine forests would recover from 30 years of intensive logging.

His research constitutes a wonderful story, hugely informed by serendipity, and provides a very clear understanding of how ponderosa pine establishes itself, and all the challenges that species face in reproducing and even surviving: Pearson determined that climate was the single greatest factor affecting its establishment, with moisture being the greatest single part. According to Pearson (1931), ponderosa pine's seeds nearly always germinated in response to summer storms, rather than in spring, because the months of May and June are the driest and warmest of the year, leading to high mortality rates even among seedlings established in the previous year's monsoon. In most years of his experiments, soil moisture in June was very low to just about nonexistent.

Pearson recognized that large numbers of seedlings became established only at long and irregular intervals, specifically when a good seed crop was followed by two or more years of favorable precipitation. Boy, did he find that out in 1919 (see below)! His long-term studies also showed that a small number of ponderosa pine seedlings established in most years with average weather conditions.

Pearson's study showed that 1919 was an 'epoch-making' year in ponderosa pine reproduction (Pearson 1923). In 1918, ponderosa pine's seed crop over the entire Colorado Plateau was unusually abundant, and the summer rainfall of 1919 was the heaviest recorded in many years, with considerable rain in both May and June, showing the importance of the timing as well as amount of precipitation. In the 13 years of his study, it rained significantly in May and June *only* in 1919. So seeds that established in 1918 were able to sail through normally dry May and June with lots of moisture! It is noteworthy that Salzer & Kipfmueller (2005) found the period between 1905 and 1922 to be the longest wet interval encountered over the last several thousand years on the Peaks, based on their study of bristlecone pines! In 1919, Pearson counted 60,000 ponderosa pine seedlings per acre on his plots! And though numbers declined to an average of 7,500 seedlings per acre by 1921, very dense forests of ponderosa pine ultimately became established from this event. I remember seeing endless doghair thickets of pine around Flagstaff, when I moved here in 1975, as remnants of this event. In many areas, they became very unhealthy forests, made up of dense spindly trees with little to no understory (Fig. 3.24). They were dark, unhealthy forests with little diversity. Doghair thickets were established through much of the West during that time (Mast et al. 1999).

This event is especially intriguing when thinking about how ponderosa pine might have spread so quickly through so much of the West at the end of the Pleistocene. With a few hundred years of lots of moisture in winter and in summer, it's easy to imagine its otherwise unimaginably rapid expansion.

The relationship of grazing to this epoch establishment event appears to have been a double-edged sword. Pearson recognized that dense grasses often limited the establishment of pine seedlings, by physically limiting their ability to get into the ground to germinate and by competing with them later for limited moisture. As mentioned above, it is widely recognized that increases in forest trees on the Peaks and through much of northern Arizona and New Mexico from the late 1800s into the mid-1900s, were facilitated by overgrazing of grasses in the forests and parklands. But Pearson, who was charged with enhancing forest recruitment, also recognized the effects of intense grazing, on the establishing seedlings themselves, and that was another matter entirely. As hard as it is to imagine, sheep and cows often fed preferentially on pine seedlings, and either killed them outright or damaged them and their future structure. Because of this, Pearson fought hard against grazing in the pine forests around the Peaks (Ashworth 1997). He did, however, recognize that moderate grazing of grasses helped ponderosa pine to establish.

Ponderosa Pine Associates

Shrubs and trees

Despite the prominence of ponderosa pine in its forests, a decent diversity of shrubs and trees is also found. Gambel oak often gains great prominence in areas with basalt outcrops, and in thinned and burned ponderosa pine stands. Five juniper species (common, Utah, one-seed, Rocky Mountain and alligator), currants (Plate 7, 8, 9), blue and red elderberry (Plate 10), Rocky Mountain maple (Plate 11), the odd pinyon, wild rose, Utah serviceberry, Fendler's buckbrush, and bitter chokecherry (Fig. 3.25) are found in pine forests in and around the Peaks (Appendix I). Shrub cinquefoil (Plate 12) graces the western slopes of the Peaks in Hart Prairie, along with Bebb willow. Small aspen groves and Gambel oak often support a greater diversity of breeding birds than surrounding pine forests (Griffis 1999, Rosenstock 1998).

Understory plants

There are more species of understory plants associated with ponderosa pine forests than even mixed conifer forests (Appendix I; Fisher & Fulé 2004). Perhaps the sheer area of ponderosa forests exerts the largest influence over such diversity. Ponderosa pine stands support many nitrogen-fixing understory species, including American vetch (Plate 13), Fendler's buckbrush, lupines (Plate 20) and locoweed, as is typically the case in volcanic mountains. Of course, grassleaf pea (*Lathyrus graminifolius*) and Nevada pea (*L. lanszwertii* v. *leucanthus*, Fig. 4.14) are two of the most important understory plants on the mountain.

In 1912, R.R. Hill found a number of grass species to be common in openings in ponderosa forests around the Peaks: mountain muhly, muttongrass, Arizona fescue, squirreltail and blue grama grass (Arnold 1950). These same species are prevalent today. He mentioned other less common species, including black dropseed, pine dropseed, Kentucky blue grass, spike muhly, deergrass, junegrass, little bluestem and sedges, which are also still around these days. Ponderosa pine is very commonly associated with blue grama grass and muttongrass (Plate 14) at its lower elevations, while at higher elevations, as precipitation exceeds about 19 inches (480 mm), blue grama is replaced by Arizona fescue (Plate 15) and mountain muhly (Plate 16), which also occur in the mixed conifer zone (Appendix I; Moir et al 1997).

Meadows in ponderosa pine stands often contain wonderful flowers including pleated gentian (Plate 17), sneezeweed (Plate 18), Parry's bellflower (Plate 19), and silver lupine (Plate 20).

Fungi

Marylou Fairweather and Edward Smith provided a list of more than 100 fungal species that occur on the Peaks, and many of them occur in the ponderosa pine zone (Appendix II).

NAU mycologist Jack States found a high diversity of fungi associated with ponderosa pine stands around the Peaks (States & Gaud 1997, States 1990). States describes the fungal communities of ponderosa pine stands in northern Arizona as highly diverse with abundant production. More than 30 species were found in ponderosa pine stands at the base of the Peaks, and eight of them accounted for more than 90% of total fungal biomass. Many are mycorrhizae, and so have a symbiotic association with ponderosa pine roots. The tree roots provide carbohydrates—from photosynthesis—to the fungi, and the fungi provide water and nutrients, with nitrogen derived from the air. They depend entirely on animals to disperse their spores. An annosus root rot is known to infect the base of old growth ponderosa pines.

Mushrooms in ponderosa pine stands 'bloom' in spring and fall, though species richness and abundance are greatest in early fall, when warm temperatures and monsoon storms really come together (States & Gaud 1997). Denser canopies, especially of mid-aged trees, support the greatest mushroom growth and greatest diversity of species. These trees have the fastest growth rates (States 1990).

Herbivores and diseases

Southwestern dwarf mistletoe is the most devastating and persistent disease associated with ponderosa pine on the Peaks and elsewhere (Table 4.1; Robert Mathiasen, personal communication 2017). It can be seen in pines along Snowbowl Road, on the west side of the mountain. This parasitic flowering plant ultimately invades ponderosa pine's vascular system and diverts both water and nutrients for its own growth, reducing its host's growth by up to 50% and often causing its mortality. This relationship is exacerbated by drought. On top of that, infected pines are far more susceptible to herbivores including bark beetles and defoliating butterflies and moths, leading to even greater tree mortality. Fortunately for ponderosa pine, many insects and birds, including Mountain Chickadees, feed heavily on mistletoe, sometimes removing up to 75% of its stems (Mooney 2006). Witches' brooms (branches deformed by mistletoe parasitism)

Table 4.1. Mistletoes occurring on the San Francisco Peaks (Robert Mathiasen, NAU, personal communication):

Arcethobium vaginatum subsp. *cryptopodum*:
Southwestern dwarf mistletoe is the most common dwarf mistletoe on ponderosa pine.

Arcethobium douglasii:
Douglas-fir dwarf mistletoe occurs on Douglas-fir and occasionally on corkbark fir.

Arcethobium microcarpum subsp. *microcarpum*:
Western spruce dwarf mistletoe occurs on Engelmann spruce in the Inner Basin.

Arcethobium microcarpum subsp. *aristatae*:
Bristlecone pine dwarf mistletoe is endemic on bristlecone pine on the Peaks and Kendrick Mtn.

support lots of wildlife species on the Peaks; most brooms are used by mammals (Garnett et al. 2004). Southwestern dwarf mistletoe, like all mistletoes, becomes widely established where trees are dense, as is the case today in so many forests in the West. And, indeed, this parasite increased in ponderosa pine stands during the 20th century.

Ponderosa pine supports many insects, birds and mammals. There are more than 350 species of insects associated with ponderosa pine through its range, and collectively they feed on all parts of it (de Groot & Turgeon 1998). More than 180 species of vertebrates are associated with ponderosa pine in the Southwest, including more than 60 species of birds, and many mammals. Pearson (1923) found that most pine seeds were devoured by animals in small cone crop years. So it's in the big seed production years that seed crops outpace herbivore consumption. Associated insects range from bark beetles to root-feeding grubs to tip moths. Birds often help in seed dispersal *and* destruction (Pearson saw Dark-eyed Juncos plucking seed coats and cotyledons from young pine seedlings). Associated rodents include squirrels, rabbits, rats, gophers, and porcupines—all of which cause damage by gnawing bark and girdling stems. Some rodents cache ponderosa pine seeds, and consequently may be more effective in dispersing them than the wind. I often find clusters of ponderosa pine seedlings emerging from caches in my yard during monsoon season.

There have been a number of significant bark beetle outbreaks in ponderosa forests in Coconino National Forest since the early 1900s. While most western conifers have only one bark beetle associated with them, ponderosa pine is attacked by multiple species of both *Ips* and *Dentroctonus* bark beetles, probably resulting from its enormous range (Table 4.2, Lynch et al. 2008). A bark beetle outbreak occurred

Table 4.2. Insect outbreak species associated with forest trees of the Peaks, Lynch et al. (2008).

Common name	Scientific name	Plant host
Pine engraver	*Ips pini*	Ponderosa pine
Arizona fivespined ips	*Ips lecontei*	Ponderosa pine
Mountain pine beetle	*Dendroctonus ponderosae*	Ponderosa pine
Roundheaded pine beetle	*Dendroctonus adjunctus*	Ponderosa pine
Western pine beetle	*Dendroctonus brevicomis*	Ponderosa pine
Western pine shoot borer	*Eucosma sonomana*	Ponderosa pine
Southwestern pine tip moth	*Rhyacionia neomexicana*	Ponderosa pine
Douglas-fir beetle	*Dendroctonus pseudotsugae*	Douglas-fir
Western spruce budworm	*Choristoneura occidentalis*	Douglas-fir, white fir, corkbark fir, Engelmann spruce, blue spruce
Spruce beetle	*Dendroctonus rufipennis*	Engelmann spruce
Spruce aphid	*Elatobium abietinum*	Engelmann spruce, blue spruce
Western balsam bark beetle	*Dryocoetes confusus*	Corkbark fir, white fir
Fir engraver	*Scolytus ventralis*	Corkbark fir, white fir
Western tent caterpillar	*Malacosoma californicum*	Aspen
Large aspen tortrix	*Choristoneura conflictana*	Aspen
Aspen bark beetle	*Procryphalus mucronatus*	Aspen
Aspen bark beetle	*Trypophloeus populi*	Aspen

in the 1930s as a result of an unusually widespread and severe lightning storm, while outbreaks in the 1950s and 2000s were tied to severe drought (Lynch et al. 2008). Ponderosa pine mortality resulting from drought and subsequent insect outbreaks occurred on more than 70,000 acres (28,320 ha) of pines in Coconino NF, between 2001 and 2004. Mortality reached 100% in some stands, especially among those with higher tree densities, at lower elevations, on dry, rocky ridges and/or with southern exposures. Defoliating insects like the southwestern tip moth and the western pine shoot borer may defoliate ponderosa pines around the Peaks, but aren't nearly as destructive as bark beetles.

Mixed Conifer Zone (MCZ)

Even in the absence of aspen, a walk through a mixed conifer stand is more interesting—with all its different species—than the forests above or below it. These stands always seem lighter than spruce-fir or ponderosa pine stands. Of course, mixed conifer forests with aspens are very lush places. It's as though there's more life there, and in some ways, there is.

The mixed conifer forest—the mixed conifer zone—supports the greatest number of conifer species on the mountain. Whittaker and Nearing (1975) found this to be the case in their study of elevation and plant diversity in the Santa Catalina Mountains in southern Arizona.

On the Peaks, limber pine, southwestern white pine, white fir, blue spruce and aspen are found throughout the zone, with ponderosa pine drifting up from its zone and Engelmann spruce and corkbark fir drifting down from theirs. These forests are different than those below or above them; they feel more open and softer. Some of this is accounted for by the large presence of aspen, and its lush understory plants. In the Southwest, the mixed conifer zone covers 5,000 mi^2 (13,000 km^2) or 1.7% of mountains. Add to this up to ~1,800 mi^2 (5,000 km^2) of aspen forests, and we're talking a large, lush ecosystem that makes its way through western mountains. Mixed conifer stands on the Peaks range between about 8,500 and 9,500 ft (2,600 to 3,170 m) on the Peaks, and occupy 14,400 acres (5,825 ha).

Aspen, according to the 2005 USGS GAP data, accounts for more than 60% of forest canopy within the MCZ, and it extends far above the zone's coniferous species (https://gapanalysis.usgs.gov/gaplandcover/). A map of this zone today, 13 years later, might show a greater presence of conifers, as fires have been suppressed for more than 100 years.

Will Moir described mixed conifer forests as the most productive forests in the San Juan Mountains in Colorado and the Southwest (Moir 1993). Mixed conifer forests may be the most productive forests on the Peaks, as well, given Pearson's finding that they get about the same amount of moisture as upper spruce-fir forests—so they are not moisture-limited, and they also have more frost-free days than the ponderosa pine zone below; they have about the same number of frost-free days as the pinyon-juniper woodlands far below (Table 3.3, Pearson 1931). For these reasons, Tom Kolb, NAU forest ecologist, regards the mixed conifer zone on this mountain as the 'sweet spot.' Perhaps this is why the mixed conifer zone supports the greatest diversity of conifer species and the greatest abundance of understory plants (see below; Fisher &

Fulé 2004). Recent studies show that even bees and flies are particularly abundant and diverse in this zone (see Mountain Insects below; McCabe et al. 2018).

Of course there are both drier and wetter or mesic conditions in the mixed conifer zone, often determined by which direction the forests are facing. Moir et al. (1997) suggested that white fir and blue spruce in mixed conifer stands were good indicators of wetter soil. I am quite struck by how even the south-facing part of this zone is pretty lush; there are places, such as in Weatherford Canyon, that have dense, green understories, and even the odd Rocky Mountain maple (Plate 11).

The mixed conifer zone has or had a more complicated fire regime than the forests above or below it (Vankat 2013). It is suggested that these forests experienced mixed severity fires, which included frequent, fast-moving ground fires and occasionally, stand-replacing fires. Accordingly, mixed conifer fire cycles are estimated between 4 and 25 years, representing a much longer cycle than that of ponderosa pine. Fire cycles of 22 years are reported for mixed conifer forests in the White Mountains (Dieterich 1983). Some Douglas-fir stands had a mean fire interval of 10 years before fire suppression began in the late 1800s (Fulé et al. 2003a, b).

Even though the Peaks' mixed conifer forests occur at higher elevations and receive more water than the ponderosa pine zone, these forests also experienced mortality due to drought in 1950s and the early 2000s, and to associated forest insect outbreaks (Kane et al. 2014). The three insects that cause the most conifer damage and mortality on the Peaks are Douglas-fir beetle, fir engraver, and a defoliator, the western spruce budworm (Table 4.2; Lynch et al. 2008). These widespread natural enemies erupted into outbreaks in drought-stricken limber pine, Douglas-fir, and white fir stands, damaging or even killing them, while aspens appeared to be succumbing more directly to the effects of drought, intense ungulate browsing and encroaching conifers (see below). Mortality levels remained high for several decades after the 1950s episode, with the fir engraver killing off many old growth true firs on steep slopes on the Peaks (Lynch et al. 2008).

Densities of all mixed conifer species have increased with fire suppression on the Peaks, as elsewhere in the West. To make matters worse, increased densities will limit water availability for trees, exacerbating drought effects, and subjecting the forests to further outbreaks of damaging insects, mistletoes and other diseases. You can see how fire kept this mountain healthy. An interesting caveat for these higher elevation forests is that pre-fire suppression densities may have been underestimated to a degree, because more moisture there causes many trees to rot, at least to the degree where their ages cannot be determined (Vankat 2013).

Douglas-fir, the widespread anomaly

Douglas-fir (*Pseudotsuga menziesii* var. *glauca*) makes up approximately one-third of mixed conifer forests on the west side of the Peaks (Fig. 4.5; McMillin et al. 2005). Douglas-fir has the widest latitudinal range of any conifer species in western North America (Hermann & Lavender 1990). It rarely occurs in large numbers in the Southwest, as opposed to in the Northwest, where it forms enormous stands, and is strongly managed and harvested for timber. A coastal variety, *Pseudotsuga menziesii*

4.5. Douglas-fir, *Pseudotsuga menziesii*, San Francisco Peaks. Courtesy of Max Licher.

var. *menziesii*) may grow over 300 ft tall (91 m), making it the 2nd tallest tree species in the world, after giant redwoods. Trees in those populations have been dated to over 1300 years old, while trees in the Southwest rarely exceed 400 years. The oldest Douglas-fir found in a study on the west side of the Peaks dated to 422 years old (Cocke et al. 2005)! It very coolly seems to prefer its kind of mid-range elevation through the western mountains. It is suggested that it really became a major feature of mixed conifer forests at the end of the Pleistocene, after having been a minor component during the Tertiary Period (Bartlein et al. 1998).

Pearson's research on the different zones showed Douglas-fir to be pretty intolerant of frost, demonstrating why it lives in the right zone, or the 'sweet spot,' on the Peaks. Douglas-fir is readily damaged by early summer frosts and by winter frosts, which are less likely in the mixed conifer zone because it has more frost-free days than the lower ponderosa pine zone, and yet gets nearly as much moisture as the spruce-fir zone (Table 3.2). Douglas-fir generally persists longest on dry sites, and is replaced by other conifers on more favorable wet sites (Guerra-de al Cruz 2001).

Douglas-fir is one of the more fire tolerant species of the mixed conifer zone, with thick, corky bark covering its trunks and even roots, and an ability to form adventitious roots or secondary roots. It is a little more tolerant of shade than many other other mixed conifer species, and is more tolerant of a variety of substrates, including calcareous soils. All of these features contribute to its impressive range in western North America.

Douglas-fir cline

Large climate differences in the coastal ranges versus the Rocky Mountains have resulted in different varieties of Douglas-fir, as in ponderosa pine (see above). The interior variety is more tolerant of drought than the coastal variety. The interior variety tends to have slower growth rates, and ultimately attains shorter height and smaller diameters (Hermann & Lavender 1990).

As with ponderosa pine, variation in latitude and elevation have resulted in genetic differences in Douglas-fir populations within the interior West (Rehfeldt et al. 2014). Genetic differences have been found in populations extending from northern Mexico to the Chuska Mountains in northern Arizona (Bressler 2003), with some divergent populations separated by as little as 600 feet (183 m) in elevation (Rehfeldt et al. 2014). Stands of Douglas-fir also become smaller and more isolated in the Southwest, often confined to small mountains or sky islands. Genetic diversity among these populations was found to be quite high, which is encouraging as the climate warms and populations become even more isolated (Bressler 2003, Moser 1999, Li & Adams 1989).

Associations

Douglas-fir dwarf mistletoe (*Arceuthobium douglasii*) is the most prevalent and damaging dwarf mistletoe in southwestern mixed conifer forests (Table 4.1; Hawksworth and Wiens 1996). It is an important natural enemy of Douglas-fir, while it also occasionally parasitizes corkbark fir, and rarely, white fir, blue spruce and Engelmann spruce (Hawksworth and Wiens 1996). Parasitism by this plant can make Douglas-fir susceptible to disease, insects and fire. Approximately 50% of Douglas-fir stands in Arizona and New Mexico are infested with it. It reduces tree growth by as much as 60%, often leading to mortality. Dwarf mistletoe's infestations have increased more than 30% in southwestern forests in the last 60 years, encouraged by higher densities of trees (Nowicki 2003).

Sometimes mistletoe infestation leads to the deformation of branches, producing what are called 'witches brooms.' Many mammals, including red squirrels, raccoons, grey-collared chipmunks, deer mice, Mexican woodrats, and porcupines, and birds including the Sharp-shinned Hawk, Common Raven, American Robin, Hermit Thrush and the Western Tanager nest in the brooms (Hedwall 2000). On the Peaks, arthropod species richness, including insects such as aphids, has been found to be twice as great in brooms as in normal branches.

The Douglas-fir beetle, a bark beetle (*Dendroctonus pseudotsugae)*, attacks and kills Douglas-fir on the Peaks. Outbreaks occurred following drought in the 1950s and its populations exploded again in response to drought in the early 2000s (Table 4.2, Lynch et al. 2008). Douglas-fir mortality was high for more than a decade following the late 1950s drought. Following the more recent drought, mortality of Douglas-fir reached more than 70% on the north side of the Peaks, and nearly 40% along Snowbowl Road (McMillin 2005)! Large numbers of dead trees were also seen on the north side of Dry Lake Hills and Little Elden Mountain. And as with other conifers, beetle attacks are greater on large trees that are also infected with Douglas-fir dwarf mistletoe. The defoliating western spruce budworm and the fir engraver also attack Doug-fir on the Peaks. So Douglas-fir, like most other forest trees mentioned, has been compromised tremendously by drought and mistletoe and insect damage. Root damage may also play a role in susceptibility to all of these forces.

White fir

Abies concolor is a beautiful conifer, with light blue-green, flattened needles; it grows to nearly 100 feet (30 m, Fig. 4.6). It has a smaller range than many western conifers, occurring from Idaho to northern Mexico. It favors mesic or wetter sites, establishing in partial shade, and decent soils. The presence of both white fir and blue spruce are good indicators of moist soils. Like so many western conifers, it is characterized by two varieties, with the coastal California white fir (*Abies concolor* var. *lowiana*) growing taller due to more moisture in the coastal mountains than in the drier interior West. Throughout its range, it exhibits great

4.6. White fir, *Abies concolor*. Courtesy of Max Licher.

variation in traits like number of needle pores or stomates, growth rates and timing of growth (Hamrick 1976).

On the Peaks, white fir is found between about 8,000 and 10,000 feet (2,400 to 3,000 m) on most sides of the mountain, and in the Inner Basin. White fir is rather uncommon on the Peaks compared with other southwestern mountains, for reasons unknown (Peter Fulé, personal communication 2017). White fir was well-represented in a mixed conifer population studied on the east side of the Peaks in the early 2000s (Heinlein et al. 2005), though that was burned by the Schultz Fire in 2010. Heinlein et al. estimated that its densities had increased dramatically since the late 1800s. Merriam didn't find white fir on the Peaks, though he did find it on the east side of Mt. Elden and north-facing slopes in Grand Canyon (Merriam 1890).

White fir is regarded as one of the more shade-tolerant western conifers, though it grows best in sunlight. It is also regarded as more fire-tolerant than many other conifers of the mixed conifer zone, due to having thicker bark. White fir is fairly intolerant of spring and fall frost, so like Douglas-fir, it lives in a zone that has a greater number of frost-free days (Table 3.3; Pearson 1931).

As with corkbark fir, white fir cones are produced high in the canopy, and typically disintegrate when their winged seeds mature, leaving behind a dark, residual spike that held the cone in place. I saw these in the tops of corkbark fir, and perhaps white fir, from the Snowbowl chairlift. At first I thought they might be some type of disease. But the disintegration of cones is why you rarely see them on the ground.

White fir and corkbark fir don't overlap often on the Peaks. Based on genetic studies, it appears that they aren't all that closely-related, either, though they are both true firs (Aguirre-Planter et al. 2011). White fir's young twigs are smooth and hairless, its needles have resin ducts near the lower end, it has grayish-green cones, and bracts in the cone scales have a short, triangular tip. Meanwhile, corkbark fir has hairy twigs, central resin ducts, purple-brown cones and bracts with a long tip (Kearney & Peebles 1964).

White fir is prone to heart rot, making it a favorite habitat for woodpeckers and other cavity nesting birds. The threatened Mexican Spotted Owl preferentially nests in old growth mixed conifer stands that are composed mainly of Douglas-fir and white fir (Zhou 1994).

Western spruce budworm (*Choristoneura occidentalis*) defoliates white fir, along with spruces, corkbark fir, and Douglas-fir, and particularly with drought. White fir is rarely attacked by Douglas-fir dwarf mistletoe.

Blue spruce

Picea pungens is typically regarded as a mixed conifer zone species, though it has also been recorded at much higher elevations on the west side of the Peaks (Fig. 4.7; SEINET.org). It is neither common nor well-understood on the mountain. Blue spruce is related to Engelmann spruce, and may have diverged from it during the Pleistocene, within the last 1 to 2 million years (Ledig et al. 2004). Though so recently diverged, they differ greatly genetically (Ledig et al. 2004). Blue spruce is widespread through the Rockies and on some peaks in the Intermountain West, though it has a smaller western distribution than Engelmann spruce. Like other western conifers, blue spruce populations exhibit genetic diversity across elevation and latitude from south to north through the West (Rehfeldt et al. 2004).

Differences in the color and texture of current year twigs are useful for distinguishing blue spruce and Engelmann spruce: the new twigs on blue spruce are often reddish in color, and glabrous or hairless, while new Engelmann spruce twigs are yellowish-tan and hairy or hirsute. Blue spruce also has a broader crown, thicker bark, spine-tipped needles and longer female cones (Springer et al. 2009).

On the Kaibab Plateau and the White Mountains, blue spruce often occurs along forest margins, while Engelmann spruce is often found in the more interior parts of forests in the mixed conifer and spruce fir zones. Their relationship on the Peaks is not understood.

Blue spruce is one of the more drought tolerant spruce in western North America (Ledig et al. 2004). Pearson's experiments showed it to be more drought tolerant than Douglas-fir on the Peaks (Pearson 1931). It is predicted, with warming, that blue spruce will be the dominant spruce on the plateaus of central and southern Utah by 2060 (Rehfeldt 2004). It may also replace Engelmann spruce on the Peaks.

Limber pine and southwestern white pine

It turns out that limber pine (*Pinus flexilis*) has a very close relative, southwestern white pine (*P. strobiformis*; Fig. 4.8), which also occurs on the Peaks, and they both have amazing stories to tell. These pines are bird pines, having co-evolved a highly synergistic relationship with Clark's Nutcracker and other jays. The trees produce very large seeds that often have no wings, and birds disperse them—and eat most of them. But they hail from Mexican and Central American white pine

4.7. Blue spruce, *Picea pungens*. Courtesy of Max Licher.

4.8. Top: Southwestern white pine, *Pinus strobiformis*; below: Left: Limber pine cone, and Right: Southwestern white pine cone.

ancestors whose seeds were winged—like most pines—and so, wind-dispersed! White pine expert Ronald Lanner (1996) says that pine seeds in the wetter tropics don't need help from birds to become established, hence the wings, but in drier regions north of Mexico seeds are more likely to germinate when buried or cached. So, while pine seed wings may help disperse pines in many climates, the caching of their seeds by birds and other animals in a drier climate can help greatly with their establishment! Partial wings have been found on southwestern white pine's seeds, leading researchers to suggest that this pine's seeds are dispersed by both birds and the wind (Kristen Waring, personal communication 2018). Limber pine seeds don't have wings. Limber pine occurs from the Peaks north through the Rockies, totally overlapping with Clark's Nutcracker's distribution; southwestern white pine occurs from the Peaks south into Mexico. It may be that both species persisted in scattered refugia during the Pleistocene, and have been spreading outward from many points ever since (Millar 1998). One of limber pine's Pleistocene refugia may have been near the Peaks (Mitton et al. 2000). These two species are thought to have diverged from one another about 18 Ma (Menon et al. 2018).

Limber pine and southwestern white pine meet and hybridize on the Peaks (Menon et al. 2018). Recent genetic studies indicate that most populations on the Peaks, including those at higher elevations, are hybrids containing >70% southwestern white pine genetics, with the remainder coming from limber pine. Populations of these pines throughout the mountains of Arizona and New Mexico are also hybrids, except for a few populations in southern Arizona (Menon et al. 2018). The two pines have been distinguished by their cones, with southwestern white pine's cone scales being recurved, or bent downward at the front (Fig. 4.8; MacDougall 1973). The discovery of hybridization between them on the Peaks makes the value of this distinguishing feature less clear for now.

Both species are recorded from pretty much throughout the mountain. They are mostly associated with the mixed conifer zone, although they occasionally extend up to treeline with Engelmann spruce and bristlecone pine, on some very harsh and exposed sites. Pearson (1931) determined that limber pine is limited by cold temperatures at higher elevations and by limited moisture at lower elevations. Like most pines, limber pine and southwestern white pine thrive in full sun, although southwestern white pine is somewhat tolerant of shade, and is reported to have greater regeneration since fire suppression began more than 100 years ago (Looney & Waring 2012, Goodrich et al. 2016).

Southwestern white pine can be found flourishing in ponderosa pine stands in Hart Prairie these days and provides a great example of a wide-ranging western pine that never seems to form large stands. The oldest limber pine or southwestern white pine found in a study on the west side of the Peaks dated to 335 years (Cocke et al. 2005). Both trees assume heights up to 100 ft (30 m) on the mountain.

Research on southwestern white pine has shown that this species, like so many other western conifers, exhibits a gradient of diversity or cline through its southwestern range, along varying latitudes and elevations (Goodrich et al. 2016). This may help it to navigate climate change, though its distribution is predicted to shrink in the days ahead (Shirk et al. 2018).

Mixed Conifer Associates

Flowering plants in mixed conifer forests

There are wonderful deciduous understory plants, shrubs and sub-trees in the mixed conifer forests on the Peaks. Shrubs and trees include Rocky Mountain maple (Plate 11), mountain ash (Plate 21), wax currant and orange gooseberry (Plates 7, 8, 9), rock spirea, and bitter chokecherry (Fig. 3.25). Rocky Mountain maple typically occurs in mesic areas such as shallow canyons and north-facing slopes, although you can see it in Weatherford Canyon. Large patches of brackenfern can be knee-high under aspen canopies (Plate 22). Arizona mountain ash (Plate 21) is recorded from the north side of the Peaks and along the Humphreys trail. I have seen it along the Waterline road on the north side of the Peaks, in deep shade under aspen and other mixed conifer trees. It has a very small, very southwestern distribution, occurring largely in mountains in Arizona and New Mexico: from the north rim of Grand Canyon and mountains near Santa Fe and Taos south to the Chiricahua Mountains in southern Arizona. Orange gooseberry, *Ribes pinetorum* (Plate 8), has the dubious distinction of being the most important vector, among *Ribes*, of the deadly white pine blister rust in the West. Limber pine and southwestern white pine are susceptible to this nonnative rust, and can be killed by it.

Bitter chokecherry, *Prunus emarginata*, is found thriving these days on the east side of Schultz Peak (Fig. 3.25). I noticed this plant a long time ago, growing along road margins to gain a little sunlight, in what was then a dense forest. It never bloomed and seemed quite marginalized. But the catastrophic Schultz fire in 2010 set it free, and since then it has greatly expanded its distribution along the slopes, as an important stabilizing force. It is clonal or spreading, just like aspen, and both are very important in holding the little soil that there is on Schultz's steep slopes (Fig. 3.25).

Flowers often found in the shady understory and along the margins of these forests include Richardson's geranium (Plate 23), greenflowered wintergreen (Plate 24), great red paintbrush (Plate 25), yellow columbine (Plate 27), Canadian white violet

(Plate 26), beautiful fleabane (Plate 28), pygmyflower rockjasmine (Plate 30), and little false Solomon's seal (Plate 29).

Herbivores and diseases

The mixed conifer zone, with its diversity of trees, abundant understory and phenomenal resources such as the big seeds from limber pine and southwestern white pine, is home to many animals. There are estimated to be 70 species of insects associated with limber pine, and 16 with southwestern white pine (de Groot & Turgeon 1998). The endangered Mexican Spotted Owl (*Strix occidentalis lucida*; Fig. 4.23) typically breeds in mature mixed conifer forests in Arizona and New Mexico (see Animals below; Ganey & Balda 1988, Zhou 1994). Black bears favor mixed conifer stands with white fir and Douglas-fir for foraging and bedding (McKelvey 2004). A number of birds, including sapsuckers, are strongly tied to aspen stands.

Limber pine and southwestern white pine may be best known from stands on the west side of the Peaks, where several NAU ornithology graduate students studied the relationships between them and a bird and a mammal during the 1980s (Benkman et al. 1984). You can find an account of this in the Animal section below. They also distinguished the two tree species in part by the more reflexed or curved back cone scales in southwestern white pine cones versus the flatter or less curved cone scales in limber pine cones (Fig. 4.8; McDougall 1973), though again, the accuracy of that trait in distinguishing the two is less clear after recent research.

Another reason for today's research on southwestern white pine and limber pine is their susceptibility to an exotic and highly invasive fungal rust, white pine blister rust, *Cronartium ribicola*. It is lethal to white pines, and limber pine, southwestern white pine and bristlecone pine are all white pines. This lethal rust came from Asia via Europe, and was first discovered in North America during an outbreak in New York in 1897. Forests of limber pine alone have been devastated in the West. In the 1990s outbreaks occurred in limber pine stands in the Sangre de Cristo Mountains in Colorado, and in southwestern white pine stands in the Sacramento Mountains in New Mexico. White pine blister rust was discovered in Arizona in the White Mountains in 2009, possibly emigrating into the region from New Mexico or California (Fairweather & Geils 2011). It is suggested that the spores of this rust may be reaching coniferous hosts in the Interior West by travelling on air currents from the Pacific Southwest (Geils et al. 2010). By 2009, the Great Basin bristlecone pine, *Pinus longaeva*, was the only white pine thought to be resistant to the rust, and Utah was the only state thought to be rust-free (Geils et al. 2010). This rust disease is one of the most important and intensively studied tree diseases in the West (Harrington & Wingfield 1998). Kristen Waring, NAU applied forest ecologist, and colleagues are monitoring for the presence of the rust on the Peaks, and conducting trials to look for genetic resistance in the pines. This disease alternates between white pines and currants or *Ribes* spp. (Plates 7, 8, 9), paintbrush species and louseworts as hosts, all of which occur on the Peaks. This rust can enter its pine host through the pores or stomates in its needles! If white pine blister rust arrives on the Peaks, limber pine, southwestern white pine *and* bristlecone pine and all of the plants and animals that depend on them—will be at great risk. For now, the white pines are safe.

Kristen Waring and colleagues are also searching for drought tolerant populations of these pines and considering where some populations might be transplanted or artificially dispersed—to higher elevations or latitudes—as the world warms. Wow, assisted migration. That is where a lot of the action is these days, and it is so lovely and compassionate, this desire to preserve some of this world.

Aspen Forests

A walk through an aspen forest is the best; it is like a walk through a beautiful garden with soft light, clean light trunks, and lush green understory with lots of flowers (Fig. 4.9, Plate 22).

4.9. Aspen stand on a Core Ridge moraine in the Inner Basin, with lush understory of Nevada pea.

Within the Peaks forests, aspen covers nearly 12,000 acres (4,900 ha), with 75% of that in mixed conifer forests and the rest in spruce-fir forests, with a little in the ponderosa pine zone (Fig. 4.1, Plate 4; John Vogel, DOI). It occurs through many of the West's mountains, and has the largest distribution of any tree species in North America. It has numerous small populations in seeps and wet canyons in the Cold Desert northeast of the Peaks, perhaps vestiges of wetter times.

You can't say enough about aspen; this plant is so kind to the Earth. In the West its stands support an inordinate number of plants and animals. You can feel it on the Peaks, when you walk from a conifer stand into an aspen grove—everything changes: there is more light, and lush vegetation under foot, with grasses, sedges and flowers. On the Peaks, understory plants are more abundant in aspen stands than in other forest types—by a lot (Fig. 4.9; Fisher & Fulé 2004). Understory plant cover was 36% higher in aspen stands than in ponderosa pine stands, which had the next highest percentage of cover. Nitrogen-fixing plants, such as lupines and sweetpeas (*Lathyrus spp.*) are widespread in aspen stands on the Peaks (Fisher and Fulé 2004). Aspen leaves are more nutritious and less acidic than low-nitrogen conifer needles, and they break down readily, recycling water and nutrients quickly, which must make a huge contribution to the mesic nature of their existence (Fisher and Binkley 2000). Aspen leaves just don't pile up like pine needles do.

Ironically, Pearson (1931) didn't find aspen soils to be any wetter than others on the Peaks. Pearson compared soil moisture in stands dominated by conifers and those dominated by aspen, and actually found that soils were often drier in aspen stands. This surprised me as aspen stands always seem more mesic or lush than neighboring stands of conifers. Pearson suggested that perhaps soil moisture was more readily depleted in aspen stands because they are highly productive and consume more water, along with their ubiquitous lush understory (Pearson 1931; p. 71).

Leiberg (1904) said it so perfectly in 1901, "The aspen stands in every case represent primary restocking after exceedingly destructive fires which wiped out most of the original coniferous growth." He saw especially heavy and extensive aspen stands on the north, west and south sides of the Peaks. You can still see these enormous stands today—especially in autumn when their leaves turn yellow: they extend in a wide band from the mixed conifer through the spruce fir zones all around the mountain. This huge ring of aspens reflects a history of big fires in the upper forests. They are still there, though conifer invasion is obvious in places (Fig. 4.10).

Aspen's seeds have exacting requirements for establishment and so it rarely reproduces by seed; instead it spreads out as a clone, sometimes becoming a large stand. But I'll bet that conditions during some of the Pleistocene and early Holocene—when the glaciers were melting—provided great

4.10. Conifers encroaching in an aspen stand.

conditions for aspen to establish by seed, and expand its range into some otherwise dry places. It must have been something. Aspen reproduced by seed—sexually—and by cloning following the 2010 Schultz Fire on the east side of the Peaks: more than 10,000 stems per acre were produced, representing a combination of clonal regeneration *and* seedling production (Fairweather 2012)! Aspen also reproduced by seed following the severe Yellowstone fire in 1988 (Fairweather 2012, Romme 1995). A study in Utah found greater genetic diversity among aspen clones that experienced frequent fires (Gardner 2013).

Clones

Aspen, with its clonality, is set free by fire (Fig. 3.25). Aspen colonizes burn sites faster than other trees, by virtue of its underground stem/root system that often survives fire to send up young shoots within a few weeks of a fire. It is surely represented by many clones as it encircles the Peaks and much of the Inner Basin, based on the different rates at which leaves turn color and their different colors and hues. Aspen may have a bigger distribution on the mountain than we know. Its ability to basically produce a large forest so quickly, speaks to a profoundly large underground existence. If it was any other species, I would be uneasy. The largest known aspen clone, 'Pando' in south-central Utah, stretches over 108 acres (44 ha), contains more than 47,000 stems, with an estimated weight of 13 million pounds (6 million kg; Mitton and Grant 1996), and it is surrounded by 59 other clones (Salmon and McAvoy 2009). Although aspen clones can be quite old, reaching into the tens of thousands of years, individual trees or trunks are thought to be rather short-lived. The oldest single aspen stem found in a study on the west side of the Peaks dated to 237 years old (Cocke et al. 2005).

Even after severe fires, aspen clones can produce thousands of new stems (Bailey & Whitham 2002). Following the 1996 Hochderffer Hills fire near the Peaks, enormous aspen regeneration occurred, to the tune of 6-25 stems per 10 ft^2 (3 m^2) within a year (Bailey & Whitham 2002). All the other tree species have come in as seeds, being blown in or carried in and cached by birds or mammals.

There have been few large fires in more than a hundred years, so you can see the dark shapes of conifers encroaching their zone, as they have forever in this endless cycle. As mentioned above, large populations of aspens established after the Inner Basin fire in 1879.

Aspen Associates

A great diversity of plants and animals lives in aspen stands; next to riparian habitat, aspen stands support a greater biodiversity than other western ecosystems, and so, aspen is often referred to as a *keystone species* (Mitton & Grant 1996, Chong et al. 2001, Worrall et al. 2013). There are hundreds or thousands of plants and animals associated with aspen across its enormous North American range. Even small aspen stands in ponderosa pine forests around the Peaks attract more birds, including migrants, primary and secondary cavity nesters than the surrounding extensive pine stands (Griffis 1999). They act as nurse plants for colonizing conifers that will always replace them. Really, it's as though a great ecological synergy happens when aspen is around.

Some of this has to do with how productive aspens are, as is true for most species in the willow family, including willow, aspen, cottonwood and poplar. They are flowering plants in a sea of conifers. They grow fast, which is a great resource. In aspen, full tree photosynthesis contributes lots of resources for growth, as aspen also photosynthesizes through its bark (Pearson & Lawrence 1958). Chlorophyll levels in both leaves and bark are highest on the south-facing section of the tree. Chlorophyll is the pigment that absorbs energy from the sun to create carbohydrates—which make a lot of the world go round! Chlorophyll levels are quite high in aspen bark through winter and early summer and decline as the leaves develop and begin to photosynthesize.

Nitrogen-fixing plants, including species of *Lathyrus* and *Vicia* (Plate 13), were found to be 2 to 25 times more abundant on aspen plots than other forests types, though they were present on nearly all forest plots (Fisher and Fulé 2004). Aspen and mixed conifer zones have very similar understory plant communities, not surprisingly, since they are mostly pretty integrated (Fisher & Fulé 2004).

Brackenfern (*Pteridium aquilinum*; Plate 22), like aspen, is clonal and adapted to fire. It colonizes or recolonizes burn sites quickly. They are both extremely widespread through North America and the Northern Hemisphere. Although there is some enthnobotanical use of this fern, it is also known to be carcinogenic, perhaps the ultimate defense against herbivores. Fern Mountain, at the base of the mountain is so named due to the large stands of brackenfern on its slopes (Fig. 2.6). Brackenfern occurs widely on the western flank of the Peaks and in the Inner Basin.

There are more than 100 kinds of insects that feed on aspen, including nine species on the Peaks whose populations sometimes erupt into major outbreaks (Table 4.2). A study in recently burned aspen stands near the Peaks found more than 30 species of insects and other arthropods on aspens (Bailey & Whitham 2002). More than 130 bird species associate with aspen stands and aspen-conifer forests, a greater diversity than in conifer stands alone (Bartos 2001). Many are insectivorous, and many nest in cavities in aspen's soft wood. And of course Red-naped and Williamson's Sapsuckers occur only in aspen stands (see Birds below, Haldeman et al. 1973). In winter, grouse and black bear feed on aspen buds. More than 50 species of mammals rely on aspen stands, including dwarf shrew, pocket gopher, porcupine, beaver, hare, rabbit, deer, bear moose, bison and elk (DeByle 1985). Even underground herbivory is intense, with pocket gophers consuming new shoots and up to 20% of aspen's belowground growth in an aspen grove in Hart Prairie (Cantor & Whitham 1989).

The bad news

Aspen on the Peaks and in much of North America is facing a convergence of pretty dire challenges to its survival, and its range is declining. Aspen cover has been declining in the West since the latter half of the 20st century (Rehfeldt et al. 2009, Worrall et al. 2013). Aspens are declining due to 1/ a lack of fire—leading to conifers overtopping them, 2/ an overabundance of Rocky Mountain elk, *Cervus elephus* (Fig. 4.24), which browse them relentlessly, and 3/ climate warming.

As a result of fire suppression on the Peaks, conifer numbers have increased, as aspen declines (Cocke et al. 2005). The cycle that exists between aspen and the conifers in relation to fire is longstanding, and was highly dynamic and very natural in the fire-prone San Francisco Peaks. Most large aspen stands on the Peaks were established by the late 1800s, following the last large fires.

Aspen on the Peaks were heavily browsed by sheep from the mid 1800s to the mid 1900s, but now are heavily and relentlessly browsed by nonnative Rocky Mountain elk populations. Rocky Mountain elk are very abundant on the Peaks, and they are browsing stands of aspens and willows to death (see Mammal section below). Elk browsing has reduced aspen regeneration since the 1960s (Lynch et al. 2008), though some USFS photos show evidence of their extensive damage as early as the 1930s.

Intense elk herbivory occurred on the east side of the Peaks after the 2010 Schultz Fire, when aspen regenerated clonally and also produced seedlings. Enormous regeneration and seedling establishment—adequate to replace the stand destroyed by the fire—were eliminated by elk (Fairweather 2012). A study in Canada showed that aspen's underground roots, or parent roots, rely quite heavily on nutrients produced by the young shoots, so root growth becomes dependent on stem growth, too; the point being that the ability of aspen's parental roots to re-create aspen stands is finite (DesRochers & Lieffers 2011). The action of elk against aspen regeneration enhances a successional shift towards coniferous forests, to the exclusion of aspen (Worrall et al. 2013).

Warming and more variable climates are stressing aspen, and with that, attacks by other natural enemies are increasing. In June 1999, an intense late frost destroyed the budding leaves that were pervasive on trees all the way around the mountain and in the Inner Basin (Fairweather 1999). Aspen on Kendrick Mountain were also damaged. Its first flush of leaves was gone, and even whole trees were killed. In response, the large aspen tortrix (*Choristoneura conflictana*) defoliated the new replacement leaves, leading Fairweather to conclude that aspen was defoliated twice in 1999, once by freezing and subsequently by insect defoliation (Fairweather 1999). This freezing trauma combined with emerging drought resulted in 'aspen decline', and its effects were felt for the next 10 years! It was evidenced by extensive crown die-back of aspens, and particularly at lower elevations in the ponderosa pine zone, at the lower end of aspen's reach (Zegler et al. 2012). Aspen throughout northern Arizona were severely defoliated by western tent caterpillars during 2004, 2005 and 2007, with this species contributing greatly to aspen decline (Fairweather et al. 2007). Other insects, including the bronze poplar borer, aspen bark beetles, poplar borers and a clear-winged moth, and a canker or fungus, attacked stressed aspens, but appeared to be secondary agents in aspen decline. A study of aspen in smaller mountains surrounding the Peaks found mortality rates of up to 50% (Zegler et al. 2012); mortality was greatest at lower elevations in mixed conifer stands and ponderosa pine stands.

The situation around the Peaks and elsewhere seems like a perfect storm swirling into place: the warming, drier world here stresses aspen, and even where it is able to establish vigorously, elk destroy it, clearing the way for dark, coniferous forests—in the midst of fire suppression. It is suggested that aspen won't survive without protection from over-browsing by elk.

Recent studies predict further loss for aspen in the southern part of its range, particularly in the United States and Mexico. They also predict an expansion of the northern part of aspen's range, which may offset losses in the south for awhile. But by 2060, it is predicted that more than 25% of its current habitat would no longer be suitable (Worrall et al. 2013).

Of course, the less intimate view might also consider that aspen's decline at the southern fringes of its range is really a long-term response to the recession of the Wisconsin glaciation (Worrall et al. 2013). Indeed, there is already evidence of rapid upslope expansion of aspen in the Rocky Mountains of western Alberta, Canada, facilitated by rising temperatures and logging disturbances. So, it's all in motion as usual, and yet, a loss of aspen constitutes an ecological catastrophe.

There are great mysteries yet to be discovered about aspen, starting with their underground roots. I really wish someone would invent a camera that could photograph the structure of their root systems. Hard to imagine our forests without aspens.

Spruce-fir or Subalpine Zone

A walk in the spruce-fir zone on the Peaks is a walk through a dark, steep, dense forest (Fig. 4.11). My personal favorite is the walk up from the Inner Basin towards Agassiz Peak, where the ground is rocky, with tree roots running everywhere with pockets of a little soil and humus, and a lot of strange mosses and lichens breaking everything down, making it seem pretty lush. There is a quiet and a mysteriousness to this place that is palpable, and gazing up through a small hole in the canopy I saw the top of Agassiz still way, way up there.

Spruce-fir forests are the highest elevation forests in North America. They are really just extensions of the great boreal forests of Alaska and Canada. In the United States, they have extended down the continent's many north-south trending mountain ranges, even migrating deep into Mexico during cold times. Their forests are easily recognized by spruce and fir's dark green and very narrow canopies (Fig. 4.11). They may be the most pristine forests because they grow on steep slopes that are difficult to access for commercial purposes.

Needless to say, the spruce-fir zone covers a smaller area than the zones below it, occupying narrowing mountain tops. On the Peaks, spruce-fir habitat covers nearly 10,600 acres (4,300 ha; Table 3.2, Fig. 4.1, Plate 4). Subalpine forests in Arizona and New Mexico cover about 1.1% of the land (Moir and Ludwig 1979), totalling approximately 3,400 mi^2 (8,900 km^2).

The spruce-fir zone on the Peaks generally occurs from about 9,500 ft (2,900 m) to treeline, and it has the shortest growing season of all the mountain's forests. The subalpine forest climate is one of heavy snowfall, extreme temperatures (-50 to 90°F, -45 to 32 °C), average temperatures at near freezing (Fig. 5.1), and fierce winds that

4.11. Spruce-fir stand on a steep slope on the west side of the Peaks.

exceed 100 mph (160 kph; Tomback 1988). Pearson measured temperatures in the spruce-fir zone between 1917-1919, and found an average summer temperature of about 51 °F (10 °C; Table 3.3; Pearson 1931).

The spruce-fir zone receives more moisture than zones below (Table 3.3, Fig. 5.5, 5.6). The extra moisture contributes to a longer fire cycle than that of the lower forests, often between 150 and 350 years (see G. Waring 2011). The moisture helps decay and recycle the forests, with the help of mosses and lichens. Spruce and fir are surely the most intolerant of fire: they have thin bark, which affords little protection from fire. It is widely thought that most fires in subalpine forests are of the stand-replacing type, rather than the cool, fast surface fires which were pretty characteristic of ponderosa pine forests, and to a lesser degree, the mixed conifer stands. The burning down of the Inner Basin spruce-fir forest and aspens in 1879 shows what these fires are about (Fig. 3.21).

Spruce-fir forests are climax forests; they may be wiped out by stand-replacing fires, and replaced initially by understory plants and often by aspens, but ultimately, they re-form dense, dark forests that may persist for hundreds of years—until the next stand-replacing fire comes along. Both spruce and fir are shade tolerant, so seedlings can establish under forest canopies, and result in multi-aged stands, from seedlings to lots of old and downed trees. They are both shallow-rooted, living on steep slopes with little soil (Fig 4.12). Pearson suggested that spruce and fir rarely send roots more than a foot deep in the spruce-fir zone. How can that work?

4.12. Soil-limited spruce-fir forests.

The recolonization of much of the Inner Basin following the late 1800s fire has been remarkably rapid; basically large stands of both conifers and aspens occur through most of the area that burned 141 years ago. While the rapid development of extensive aspen stands is not surprising, the rapid return of the conifers is (Fig. 3.21). Spruce-fir forests are thought to take hundreds of years to replace themselves. It turns out, though, that the USFS replanted Engelmann spruce in the Inner Basin around 1911. The seedlings were grown from spruce seeds collected in the Inner Basin. So, this certainly helped with its speedy recovery!

Numerous avalanche chutes transect the spruce-fir zone on the outer flanks of the Peaks and in the Inner Basin, and represent a significant disturbance in this zone (Fig. 3.6; see Glaciers and Avalanches above, KachinaPeaks.org). Some avalanches are powerful enough to scour great swaths of spruce and fir (Fig. 3.7). I've seen this in Snowslide Canyon, the canyon loaded with snow and shredded spruce and fir trees—in the month of June! Many trees are killed or damaged, though studies have also shown that the trees that survive often grow more, due to less competition and more light (Dexter 1981). Lee Dexter, NAU Geographer, aged the spruce and fir that were ripped out by the enormous 1973 Dunnam Canyon avalanche and concluded that that particular path had about a 100 year return cycle (Dexter 1981). There are also occasional blow downs from strong winds, which can predispose damaged trees to insect damage (See Associates below).

The Peaks' spruce-fir forests are relatively simple: they are the essential boreal extension. Engelmann spruce (*Picea engelmannii*) and corkbark fir (*Abies lasiocarpa* var. *arizonica*) dominate the forests (Fig. 4.13). Engelmann spruce occurs through the upper spruce-fir zone, even occurring in a stunted form at treeline. Corkbark fir often dominates the lower part of the zone. Engelmann spruce and corkbark fir are distinguished readily by cones (Fig. 4.2). Engelmann cones are persistent, and seen hanging downward in the upper canopies and clustered on the ground. Fir cones stand erect on upper branches, and fall apart when they ripen, leaving behind only a dark spike that the cone scales formed around. Spruce has stiffer and sharper needles, and its bark is rather reddish-brown and flakes off in little tile-like sheets. Corkbark fir bark is corky looking and light gray (Fig. 4.13).

Spruce-fir stands are more open on drier southern and western exposures, with limber pine and bristlecone pine often joining in.

Engelmann spruce

Engelmann spruce is more common and longer lived than corkbark fir. It reaches heights of 80 to 100 ft (24 to 30 m) and 400 to 500 years in age (Fig. 4.13). The oldest Engelmann spruce found in a study on the west side of the Peaks dated to 320 years old (Cocke et al. 2005). Both species are slow growing, and able to tolerate long-term shady conditions, though spruce also colonizes open sites. Pearson (1931) observed 'overstocked' stands of spruce and fir on the Peaks, with increased densities a result of fire suppression. In such conditions trees were typically smaller, rarely exceeding 80 ft (24 m) in height, than when occurring in lower densities, where free-standing trees reached 100 ft (30 m).

Engelmann spruce clines

Engelmann spruce is one of the most widespread North American spruces, occurring from British Columbia through Arizona and New Mexico. Engelmann spruce is a complex of diverse populations distributed along a south to north gradient or cline dictated by intensifying winter temperature regimes (Rehfeldt 2004). Gerald

4.13. Left: Corkbark fir, *Abies lasiocarpa* var. *arizonica*. Right: Engelmann spruce, *Picea engelmannii* in the Inner Basin.

Rehfeldt described a northern population of Engelmann spruce which hybridizes with boreal white spruce, Engelmann populations through the middle Rockies, and smaller southern populations. Engelmann spruce has large and extensive populations in the northern part of its range and then by Arizona and New Mexico occurs only in smaller and increasingly isolated mountains or sky islands such as the Peaks. Its southernmost population occurs in the Chiricahua Mountains near the Mexican border. Spruce extending into Mexico, through the Sierra Madre Occidental and the Sierra Madre Oriental, are clearly related to Engelmann spruce, though they have diverged considerably since the Pleistocene. Their story tells what happens to populations that spread out when the climate is favorable or cool, and then get stuck in what become less favorable conditions as climate warms, as since the end of the Pleistocene. Once wide-spread populations in Mexico have become increasingly isolated among Mexico's many ranges that are surrounded by desert. Some populations are already gone, others are shrinking, and yet one spruce population is unique and may be one of the more drought tolerant spruce populations known (Ledig et al. 2004)!

The climate gradients of mountain islands that Engelmann spruce grows along in the Southwest are much steeper than those of populations further north in the Rockies and in the Canadian boreal forests. According to Rehfeldt (2004), based on global warming predictions, the climate of southern Arizona and New Mexico will no longer be suitable for existing Engelmann spruce and blue spruce by the year 2090, with populations on the Peaks and the White Mountains becoming isolated by 2030. Shortly thereafter, blue spruce may come to dominate mountain forest islands in Utah, and perhaps even areas in northern Arizona that are currently dominated by Engelmann.

Corkbark fir

Corkbark fir is endemic to the southern Rocky Mountains and smaller ranges in Arizona and New Mexico, and has yet to be adequately studied (Fig. 4.13). Surely, it has the smallest distribution of the western conifers, occurring only from southern Arizona and New Mexico to southern Colorado! During his initial survey in 1889, C. Hart Merriam described it as having a 'fine elastic cork' bark and short needles, and occurring mainly in the lower part of the spruce-fir zone (Merriam 1890). He suggested that it was distinct from its parent species, subalpine fir, which occurs from the Rockies to Canada, but didn't formalize that speculation, due to an absence of material with which to make comparisons. Merriam returned to the Peaks in 1896, and collected the type specimen for this species now described as *Abies lasiocarpa* var. *arizonica* Merriam (see SEINET.org). Plant taxonomists Kearney & Peebles said that there are no other constant physical characters that distinguish corkbark fir from its widespread parent, subalpine fir (Kearney & Peebles 1964). And yet, corkbark's bark is really different from its parent subalpine fir: it has thick, corky and creamy white bark, while subalpine fir has thin, grayish-brown bark. More recently, a DNA comparison of essential oils in the leaves of subalpine fir and corkbark fir showed them to be very different (Adams et al. 2011). I think a study of the ecology of this southern-dwelling conifer—with its weird, thick bark, and its relationship to subalpine fir, would be the coolest thing in the world. It seems likely that corkbark's thicker bark must serve it well and enable it to live in the drier Southwest, basically extending the range of subalpine fir into the Southwest.

Corkbark fir rarely lives longer than 250 years. The oldest corkbark fir found in a study on the west side of the Peaks dated to 214 years old (Cocke et al. 2005).

Spruce-fir Associates

One of the most comprehensive plant studies on the Peaks is that of Laurie Ann Paulik, NAU Master's student, who in 1980 described the species associated with the spruce-fir zone. She also compared the understory plants of the spruce-fir zone with those of the tundra and the lower zones. She was thinking about elevation and plant distributions, too.

Paulik's study found 189 plant species in the subalpine zone (Paulik 1980). More than 250 are known from there today (Appendix I). She found, remarkably, that only 5% of the understory plants found in the Peaks' spruce-fir zone occurred *only* there, including Eschscholtz's buttercup, snowline wintergreen and hoary ragwort (Appendix I; Paulik 1980). Rather, more than 30% of the understory species were shared with the tundra zone above, and 60% are found in the mixed conifer zone and sometimes even ponderosa pine zones below.

Paulik (1980) described several different communities of understory plants within the spruce-fir zone. The densest spruce-fir forests have little understory cover. Nitrogen-fixing legumes, including American vetch (Plate 13) and Nevada pea (Fig. 4.14) make up very lush understories in more open stands in the lower spruce-fir zone, as well as in mixed conifer stands (Fig. 4.9). And then there's the loveliest orchid, the hot pink fairy slipper (*Calypso bulbosa*, Plate 31) that lives along forest edges and in canyons. I saw them tucked in under spruce and fir bordering a meadow in the Inner Basin. There are diverse understory communities in canyons and ravines in the spruce-fir zone.

Meadows through the spruce-fir zone are dominated by grasses such as Arizona fescue (Plate 15), sheep fescue, mountain muhly (Plate 16), muttongrass (Plate 14), bromes, pine dropseed, Kentucky bluegrass, and sedges, and a diverse mix of flowers (Appendix I; Paulik 1980). This is the case in the Inner Basin and on slopes between Agassiz and Humphreys.

At the top of the zone, near treeline, bristlecone pine joins Engelmann spruce in more open stands, where they support one of the richest understory communities

4.14. Nevada pea, *Lathyrus lanszwertii* var. *leucanthus*, a very important understory species on much of the Peaks, and on many other western volcanic mountains.

of all, with high elevation species such as Franciscan bluebells, Whipple's penstemon (Plate 33), and sneezeweed (Plate 18) present (Fisher & Fulé 2004). This can be seen on the upper west and southwest slopes of Agassiz Peak and similar exposures elsewhere.

At least 15 species of ferns occur in the spruce-fir zone, including species of *Botrychium* which reach their best development in Inner Basin meadows, while the circumboreal species brittle bladderfern and brackenfern (Plate 22) are common, with the latter especially associated with aspen stands (Appendix I, Paulik 1980).

Most shrubs occur in the lower part of the spruce-fir zone and extend down into the mixed conifer zone. Shrubs include red elderberry (Plate 10), rock spirea, snowberry, mountain ash (Plate 21), and wild raspberry (Appendix I). Common juniper and gooseberry currant (Plate 8) occur throughout the zone. Both of these are easily seen in the meadows in the upper Inner Basin, along with twinberry honeysuckle (Plate 32). The currants are many feet in diameter in that open site, while honeysuckle is relentlessly browsed by elk.

Corkbark fir, fortunately, is only occasionally parasitized by mistletoe on the Peaks, specifically by Douglas-fir dwarf mistletoe (Table 4.1; Robert Mathiasen, personal communication). Engelmann spruce in the Inner Basin is parasitized by the same endemic subspecies of mistletoe that attacks bristlecone pine (see below).

In keeping with the greater moisture levels in the spruce-fir zone, the most common diseases in spruce-fir stands are caused by wood-rotting fungi, which can predispose trees to windthrow.

According to Furniss & Carolin (1977), there are more than 60 species of insects associated with Engelmann spruce, and they use all of its parts, from roots through stems and foliage and seeds. Two of them have been studied on the Peaks (Table 4.2; Lynch et al. 2008). Engelmann spruce is attacked by the spruce beetle, *Dendroctonus rufipennis*, a bark beetle, with population outbreaks in response to local disturbances such as windthrow, avalanches, and development at Snowbowl Ski Area. The spruce beetle is regarded as Engelmann spruce's most serious herbivore. Small, localized outbreaks occurred at the ski area in 1953-54 and in 1962-63 in association with ski trail construction, in 1967-68 on the west slope of Agassiz Peak in association with heavy ice and snow and a windthrow event, in 1973-75 in the Inner Basin following an avalanche, and near Snowbowl in 1974-76 after a high wind event (Lynch et al. 2008, McMillin 2006). A cyclonic wind event in 1999 caused windthrow of large spruce on nearly 30 acres (12 ha), and ultimately led to a spruce beetle outbreak that killed nearly 1200 spruce trees near Snowbowl, lasting until 2005. The exotic spruce aphid (*Elatobium abietinum*) established on the Peaks around 1999. Hard to imagine, but it feeds on tree resin through needle stomates or pores, damaging Engelmann spruce and blue spruce in all size classes.

According to Furniss & Carolin (1977), there may be as many as 30 insect species associated with various parts of subalpine fir. Two of its most damaging insects are associated with corkbark fir on the Peaks: fir engraver (*Scolytus ventralis*)—a bark beetle, and western balsam bark beetle (*Dryocoetes confusus*). Like other conifers on the Peaks, corkbark fir was hit hard by herbivores following the droughts of the 1950s and the early 2000s (Table 4.2; Lynch et al. 2008).

4.15. Bristlecone pine, *Pinus aristata*, at treeline, in the Inner Basin and on Hart Prairie.

Outbreaks of both western balsam bark beetle and fir engraver led to considerable corkbark fir mortality particularly on the north and west sides of the mountain. Their outbreaks are typically associated with disturbance events. Sadly, a Forest Service study in 2005 found that up to 60% of corkbark fir on the north side of the Peaks were killed by these forest insects (McMillin 2005). Windthrow, avalanches and even drought can trigger beetle outbreaks by stressing trees. Lots to contend with.

Bristlecone pine forests

Walking in a bristlecone pine stand typically requires scrambling up a steep, rocky slope towards treeline, accompanied by intense wind, cold and sun (Fig. 4.15). Although presented separately here, I consider bristlecone pine populations at treeline as occupying the upper part of the spruce-fir zone; it is a really interesting place. It occurs on about 450 acres (182 ha) there. Gustav Pearson suggested that high winds even more than low temperatures may determine treeline elevations on the Peaks.

Bristlecone, *Pinus aristata*, is a white pine, like limber pine and southwestern white pine—though it is not a bird pine, as its seeds are much smaller and they have wings! Like ponderosa pine, which also has wind-dispersed seeds, its cones have long, sharp prickles or spines that challenge herbivores after its seeds (Fig. 4.2).

In Arizona, bristlecone pine occurs only on the San Francisco Peaks, and otherwise is best represented in the Rockies. Its nearest relatives may be up to 300 miles (480 km) to the northeast, in the Sangre de Cristo Mountains in northern New Mexico. A sister species, Great Basin bristlecone pine, occurs through the Great Basin. On the Peaks, as elsewhere in the West, bristlecone pine is most commonly described as a treeline species. Merriam described it as 'a tree of high altitudes,' vying with Engelmann spruce for the highest elevations on rocky outcrops on the Peaks (Merriam 1890). High up on the Peaks, bristlecone pine generally occurs alone on rocky, exposed, sunnier sites adjacent to mountain forests. It is known from the north, south and west slopes of the Peaks, and on Schultz Peak, reaching elevations of 12,000 ft (3,600 m). You can see it on the hike to Humphreys, near the saddle it is low-growing and kind of sculpted by the wind (Fig. 4.15). In the Inner Basin good-sized stands are found on Core Ridge and south Core Ridge on rocky outcrops well below treeline. Bristlecone pine occupies such habitats throughout the West. Like most pines, it is intolerant of shade.

It is thought that bristlecone pine used to live only near treeline on the Peaks. That's never quite made sense in a way, but who knows. It is actually pretty fire-tolerant, with its thick bark, making it curious that it resides largely above the fires, on rocky outcrops at treeline. Some treeline trees have fire scars, suggesting that bristlecone could conceivably tolerate fires in the ponderosa pine zone—where it is thriving today. It is thriving at about 3,500 feet downslope in Hart Prairie, at about 8,500 ft (2,615 m; Fig. 4.15). I believe this gives bristlecone pine a greater elevational range in the mountain than the other forest trees.

A study dating its tree rings from live and dead trees on the upper slopes of Humphreys and Agassiz found individual plants over 1,300 years old, while cross dating with dead trees demonstrated its presence there for more than 2,300 years (Salzer & Kipfmueller 2005)! Another study on the west side of the Peaks dated an individual at 479 years (Cocke et al. 2005).

The rarity of bristlecone pine in Arizona has resulted in a designation as a salvage restricted protected species by the Arizona Department of Agriculture (Scott 2008).

Bristlecone pine associates

At treeline, a mix of bristlecone pine, Engelmann spruce, common juniper and understory species from zones below and the tundra above converge to make for one of the richest understory communities of the zones (Fisher & Fulé 2004, Schaack 1983). Schaack (1983) described this treeline area, the krummholz, as colonized largely by populations of stunted Engelmann spruce and bristlecone pine. Its elevation fluctuates around 11,500 feet (3505 m), depending on aspect and substrate stability. These treeline forests are more open, and yet provide critical protection from the elements for these tiny plants. Understory species there include meadow-rue, autumn dwarf gentian, fleabane daisies, diamondleaf saxifrage, and sneezeweed (Plate 18). Grasses and sedges include muttongrass, Kentucky bluegrass, fringed brome, alpine timothy, and Idaho bentgrass. Sometimes mountain parsley, bluebells, sedums, Whipple's penstemon (Plate 33), goldenrods and gooseberry currant (Plate 9) join in.

There is one area on the Peaks where bristlecone isn't doing so well, and that's on the top of Schultz Peak, at about 10,000 ft (3,000 m) (Fig. 2.6; Scott 2008). The bristlecone there have been hit very hard by a mistletoe that turns out to be endemic to the Peaks and Kendrick Mountain, bristlecone pine dwarf mistletoe, *Arceuthobium microcarpum* subsp. *aristatae* (Fig. 4.16; Scott 2008). In 2008, 95% of all bristlecone in the Schultz Peak population were found to be infected with this mistletoe and 57% were dead. Severely parasitized living trees had lower growth rates (Scott 2008). The years with the most mortality were linked to high mountain pine beetle activity and to drought (Scott 2008). This mistletoe also attacks blue spruce and Engelmann spruce, and rarely corkbark fir and southwestern white pine. Apparently, white fir, ponderosa pine and Douglas-fir are immune to it.

There are 18 species of insects associated with bristlecone pine (de Groot & Turgeon 1998). On the Peaks, one of them—a bark beetle called mountain pine beetle—attacks trees that are affected by drought and by mistletoe parasitism (Table 4.2).

Sky Island Tundra—World Above The Trees

Walking through the alpine tundra on the Peaks means walking above treeline. Conditions range from scrambling up steep scree slopes with small patches of low-growing flowers and grasses, with wind, cold and glaring sun at your back to walking through quieter, more level and wetter sites with abundant growth, often next to melting snowbanks that persist through most of the summer—all in a world too harsh for trees (Fig. 4.17). To really appreciate the tundra plants, you have to get down close to the ground, as few of them ever reach even six inches in height.

The Peaks is the only mountain in Arizona that supports tundra habitat, by having multiple peaks that reach above 11,500 feet—above treeline (3,500 m). This zone covers nearly 1,200 acres (485 ha; Table 3.2). The tundra extends up the highest peaks: Humphreys, Agassiz and Fremont Peaks (Fig. 2.6, 4.1, Plate 4). It's amazing that these plants are here. The zone above treeline on the Peaks is harsher than tundra habitats in the Arctic, causing William Billings, alpine ecologist, to describe tundra habitats in the western United States as essentially high elevation deserts (Billings 1974). That may be especially true on the Peaks, being at latitude 35° N and all, up to 12,633 ft (3,850 m)—where wind, temperature and UV rays are incredibly intense. Remember that tundra occurs at elevations as low as 300 ft (91 m) in the Arctic, while increasing in elevation onto mountaintops towards the equator.

What luck: I believe the only published temperature measures for the Peaks' tundra zone are from the top of Agassiz, from a study there in August 1898. Two scientists with the Carnegie Institute of Science, studying the southwestern deserts, decided to spend several weeks measuring temperature at the top of Agassiz

4.16. Left: Bristlecone pine branch infected with the endemic bristlecone pine dwarf mistletoe, *Arceuthobium microcarpum* subsp. *aristatae*, and Right: a bristlecone population on top of Schultz Peak with numerous trees killed by this mistletoe. Courtesy of Robert Mathiasen, NAU.

4.17. Tundra habitats, including Left: a fellfield or scree area, and Right: a meadow with melting snow above treeline.

Peak (Coville & MacDougal 1903)! They measured temperature at 12,356 ft (3,770 km), and shed light on why living low to the ground above treeline is the way to go. First, they found that air temperatures above treeline did not vary as much daily as they did at lower elevations, it's just plain cold above treeline. Then they found that the ground was much warmer than the air temperature, and that this pattern persisted through the night: 8 Aug. 4 pm: soil temperature = 71.6 °F (22 °C), air temperature: 57.6 °F (14.2 °C); and at 7 am the next morning, soil temperature was 48.2 °F (9 °C), air temperature was 21.2 °F (-6 °C). These patterns in temperature explain how tundra plants catch precious heat by staying low to the ground and why they are often spreading plants.

Merriam, a bit melodramatically, characterized the tundra zone as the 'bleak and storm-beaten summit' of the Peaks in 1889. It can feel that way, I suppose, with a storm approaching, but when the sun is out, it is vast, and fresh and beautiful with flowers everywhere. The tundra extends between about 11,500 ft (3,500 m) and the top of Humphreys at 12,633 ft (3,850 m). It's an edgier world though; you can feel it. Tundra habitat occurs mostly along the slopes of Agassiz, Humphreys and Fremont Peaks which reach above 11,500 ft (Fig. 2.6), making it the smallest habitat on the mountain (Table 3.2).

There have been a number of studies of the Peaks' tundra and with each the number of tundra species found has increased. C. Hart Merriam and his team reported 20 tundra species on the Peaks in the late 1800s, Little (1941) reported 49, and Moore found 51 (1965). Clark Schaack's study expanded the number of plant species above treeline to 80 (1970, 1983). Botanists at the Deaver Herbarium at NAU expanded the list to 94 tundra species in 2006. There are nearly 200 tundra species in the Rockies, showing how diverse this extreme ecosystem can be. There are fewer species on the Peaks, due in great part to its limited habitat, low latitude and great isolation. Schaack suggests that the recent age of the Peaks also explains its having fewer tundra species. He spent several seasons studying this zone.

To date, 96 plant species have been found in the tundra zone on the Peaks (Appendix I). They are mostly perennial plants, many are spreading species, and puffy or cushion plants and they are beautiful. There are many different kinds and colors of flowers, extraordinary flowers in yellows, blues, purples, hot pinks, deep pink. They are often organized in little gardens, in clusters in little terraces across steep scree slopes. In one spot, I found Ross's avens, moss campion, red-flowered mountain parsley and sticky polemonium interwoven in a little garden on a stable patch in a scree slope on Agassiz. These lovelies are also found in high elevation meadows and in cracks in boulders near the summit (Fig. 4.18, 4.19, Plates 35 to 44).

To my way of thinking, it is with the alpine tundra that this mountain becomes a full-on sky island. This is because many tundra species on the Peaks are separated

4.18. Some of the Peaks' tundra species: Top left to right: 1. sticky polemonium, *Polemonium viscosum*, 2. Ross' avens, *Geum rossii*, 3. Parry's primrose, *Primula parryi*, Bottom left to right: 4. Whipple's penstemon, *Penstemon whippleanus*, 5. spreading wheatgrass, *Elymus scribneri*, and 6. moss campion, *Silene acaulis*. *Geum rossii*, *Elymus scribneri*, *Silene acaulis*, and *Primula parryi* photos courtesy of Max Licher.

by hundreds and hundreds of inhospitable miles from tundra habitats and populations of the same species. Clark Schaack (1983) estimated that 60% of the tundra plants on the Peaks are from the Rockies, nearly 250 miles away (400 km), while a whopping 25% are from the Arctic, more than 2,000 miles away (3,200 km)! Many of the latter have a circumpolar distribution, meaning that they occur through much of the northern part of the world (Schaack 1970).

There are about about 14 plant species that occur *only* in the tundra on the Peaks. The Peaks is also the only place they are found in Arizona, with none occurring in the White Mountains. They include several sedges (Plate 34), black-seeded dandelion, northern buttercup, whiplash saxifrage, tufted alpine saxifrage, and snow cinquefoil (Appendix I). Their populations are especially vulnerable to climate change.

The ability of these delicate, low plants to get to the Peaks, perhaps migrating through deserts in places, is in some ways the most profound aspect of the Peak's story. It all has to do with the last ice age, the Pleistocene (see Section 3). The numerous islands of tundra communities through the Rockies were larger and more interconnected during the Pleistocene glacial phases, making it a little easier to imagine how these tiny plants migrated from the Arctic to the Peaks. Certainly Merriam thought about it, and that they had made their way here during the Pleistocene cold. Clark Schaack's tundra study suggested several migration routes along which the tundra species made it to the Peaks: including down through the Rockies through the White Mountains and across the Mogollon Rim (Schaack 1970). It has been suggested that the Mogollon Rim was suitable for tundra plants during the last Pleistocene glacial (Hevly 1964). There may have been other routes, too (Fig. 3.10). Schaack suggests that some plants that range from lower elevations into the tundra may have migrated in from elsewhere on the Colorado Plateau and even across the cold desert of the Navajo Reservation to the Peaks. There are also many mountain plants that occur on both sides of the Grand Canyon, and tundra plants were there during the last glacial (Weng & Jackson 1999), so it seems likely that plants somehow managed to cross *that* divide, too. Long-distance migration of seeds via wind and birds is often suggested to explain some migrations, and yet, as Schaack points out, the many possible migrational corridors eliminate this as a necessity for most species that arrived at the mountain, and given the small size and isolation of the Peaks, it would have made for an improbable target for most seeds being dispersed by wind or birds.

Though southern New Mexico and northern Arizona are the end of the line for tundra species in the Southwest these days, there are tundra species in mountains in Mexico and Guatemala that also came from the Rockies, while tundra communities even further south are more closely related to those in the Andes! These tiny, poor dispersers certainly do get around.

A remarkable aspect of tundra plants is that many are very long-lived, with some cushion plants living hundreds of years, while some clonal species may live for thousands of years (see Waring 2011). A study of moss campion in Alaska found that large cushions lived an average of 338 years, with some living substantially longer (Morris & Doak 1998). They are also slow growing, given low temperatures and a very short growing season. Along with slow growth

4.19. Pink stonecrop, *Rhodiola rhodantha*, lives in crevasses above treeline.

and limited growth opportunity, alpine plants have some of the highest root to shoot ratios of any plants, with up to 90% of their biomass residing in their roots (see Waring 2011). This store of nutrients sees them through prolonged cold and snowy times.

Tundra habitats on the mountain include fellfields or slopes of scree and alpine meadows, and a few rockier, boulder fellfields (Fig 4.17). Plants also live in cracks in rock faces, creating the loveliest little gardens (Fig. 4.19).

Fellfields dominate the steep exposed slopes on the Peaks, with steeper slopes supporting fewer plants (Fig. 4.17). Slopes can exceed 30°. These are the most severe habitats, with little stability, dry and poor soils, and enormous exposure. Despite this, Ross' avens (*Geum rossii*; Fig. 4.18, Plate 35), is most common in the fellfields, as well as most of the alpine meadows (see below). Species restricted to or found predominantly in the fellfields include ferns, pussytoes, alum root, moss campion (Plate 36), Tweedy's sandwort, sticky polemonium (Plate 37), San Francisco Peaks ragwort (Plate 38), and spreading wheatgrass (Fig. 4.18) (Appendix I; Schaack 1983). Little (1941) noted the commonness of the alpine fern, *Cystopteris fragilis*, in crevices in the fellfields, regarding it as the only species of fern there. Today three species of ferns are known from treeline (Appendix I).

Boulder fields are found on the southwest side of Agassiz Peak. Large overlapping boulders there create protected areas for plant growth. These may also include rock avalanche chutes around the mountain. Plants there include ferns, bluebells, alum root, fireweed, Ross' avens (Plate 35), and shrubs including gooseberry currant (Plate 9), and twinberry honeysuckle (Plate 32).

The alpine meadows contain the most tundra species. Some of the best developed meadows are on the south face of Agassiz Peak, due largely to lots of sunshine and moisture and sometimes more protection, and at the top of the Inner Basin in what has been called the 'Humphreys cirque group,' a cirque left by all three known glaciers (see glaciers, Fig. 4.17, Fig. 3.2; Little 1941). The latter are protected from westerly winds, being on the lee side of the Humphreys saddle, and are often fed water from melting snowfields into late summer, so it is very wet, with enormous fields of Ross's avens, sedges and the queen of the high elevation flowers, Parry's primrose (Fig. 4.17, 4.18, Plate 40). Other plants that occur in alpine meadows include ferns, bromes, muttongrass (Plate 14), Kentucky bluegrass, sedges, an endemic buttercup, *Ranunculus inamoenus* subsp. *subaffinis* (Plate 39), creeping sibbaldia, gentians including annual gentian, *Gentianella tenella* (Plate 41), perennial fringed gentian, *Gentianopsis barbellata*

(Plate 42), Parry's lousewort, American alpine speedwell, and sneezeweed (Plate 18). Both spreading wheatgrass (Fig. 4.18) and spiked woodrush (Plate 43), along with most plants, are found in both fellfields and meadows (Schaack 1983).

Plants found in cracks and crevices in boulders include ferns like Reeve's bladderfern, pink alumroot, tufted alpine saxifrage, and pink stonecrop (Fig. 4.19), and at higher elevations on Agassiz, alpine mountain-sorrel (Plate 44).

Schaack (1983) found that even above treeline, plants could become water-limited, observing that blooming of late-flowering species was 'drastically curtailed' in years when snowfall was light or near absent, and summer storms infrequent. He found that Whipple's penstemon bloomed with good monsoon rains, and didn't bloom in their absence. It has been suggested that higher elevation plants are somewhat insulated against drought, and yet these findings suggest that low precipitation limits plants at high elevations, too.

The Species

The San Francisco Peaks ragwort or groundsel, ***Packera franciscana***, is endemic to the Peaks, meaning that it occurs nowhere else. It is small, rarely reaching even 6 inches tall, a small perennial plant that spreads, and has lovely bright yellow flowers (Plate 38). While other tundra species might face extirpation or local extinction with climate changes leading to habitat loss, if the Peaks ragwort loses out on the Peaks, It. Is. Gone. As a single-mountain endemic associated with this high elevation, sky island, its existence is particularly precarious. In 1987, the U.S. Fish and Wildlife Service listed it as threatened, and developed a plan for managing its habitat and monitoring its condition. It is a remarkable species, living happily in some pretty large populations on scree slopes on Agassiz Peak, especially where conditions are extremely harsh and unstable. This species is highly adapted to the dynamic nature of scree habitats and disturbance, and its adaptations include low stature and a reproductive system that relies on both vegetative spreading and sexual reproduction with its prolific flowers (Plate 38). When frost or rockslides break up clones, they are able to resprout and expand. However, one study did find evidence of dead individuals at the base of some avalanche chutes (Fowler & Seig 2010). This species was studied along several trails leading up to the Humphreys saddle. It is nearly impossible to depart much from the trail without potentially destroying plants in this incredibly fragile habitat. They found more than 36,000 individual plants along the Humphreys trail and the upper Weatherford Trail, and in one year, nearly 10% of the plants were either in bloom or setting seed! This study suggested that they are doing well in their Peaks home.

In the 1990s, a population of Peaks ragwort with unusual leaf morphology was found near Snowslide Spring in the Inner Basin (Johnson & Ayers 2013). It turned out that the Peaks ragwort was hybridizing with a closely-related species, hoary groundsel or *Packera werneriifolia*, that typically occurs at slightly lower elevations on the Peaks, surrounding local mountains, and further north in the Rockies (Appendix I). Perhaps the Peaks ragwort is an offshoot of this widespread species. Their hybridization was surprising since the two species typically bloom at different times, with *P. werneriifolia* blooming earlier in the summer. Researchers found out that Peaks ragwort populations are most strongly represented on Agassiz Peak and in Snowslide Canyon, where there is some hybridization, while another Inner Basin population was more represented by *P. werneriifolia*, with lower levels of hybridization. Even hybrid populations in Abineau Canyon on the north side of the Peaks are largely made up of *P. werneriifolia*. On Kendrick Mountain several miles to the west, populations consist almost entirely of *P. werneriifolia*. So, what does this mean? Perhaps these species are, through hybridization, imbuing one another with the right mix of genes to expand into new habitats, and to tolerate changing climates. Could *Packera franciscana* ultimately live on the Peaks through hybridization with the lower elevation *P. werneriifolia* as the climate warms? I sure hope so.

- ***Ranunculus inamoenus var. subaffinis,*** the endemic graceful buttercup, was collected on the Peaks at 12,000 ft (3,600 m) on Agassiz Peak in 1884 (Plate 39; Kearney & Peebles 1964). It was recognized as different from its parent, *Ranunculus inamoenus*, which is widespread through the Rockies. It has been suggested that "it differs from the Colorado and northern [populations] only in its more slender habit, minute flowers, and long cylindrical achenes" (Benson 1948), so the differences are subtle. The type specimen was collected near the top of Agassiz Peak (Benson 1948). Paulik's study found this endemic in both the spruce-fir and the tundra zones, and in the treeline community between them (Paulik 1980). She determined that the range of this species on the Peaks extended from Snowbowl up Agassiz east to Fremont Peak, north to Abineau Canyon, and in the Inner Basin. Schaack (1983) found this endemic to be common in alpine meadows. It has even been found at much lower elevations in some of the foothills of the Peaks, including White Horse Hills. This greater elevational range may serve this restricted species well in the face of global warming.

- Ross' avens, ***Geum rossii***, is the backbone of the tundra community on the Peaks (Merriam 1890, Little 1941, Moore 1965, Schaack 1983). This perennial rose is widespread above treeline through the Rockies, but in Arizona it occurs only on the Peaks. On the Peaks, it occurs abundantly throughout the tundra—in fellfields and meadows—always as a dominant plant, with its spreading or clonal clusters of leaves and long-stemmed yellow flowers. It forms little terraces through the fellfields (Fig. 4.18, Plate 35). Its large populations on steep, scree slopes above treeline surely help to hold those slopes in place. Ross' avens is also present where treeline meets the tundra, often forming large mats there along with many other tundra species. A Montana study of Ross' avens in recently disturbed sites and in more stable, established sites found that plants expanding within disturbed sites grew more and reproduced more than those in more established tundra communities, showing that it can thrive in different settings above treeline (Chambers 1991).

Ross' avens is easy to spot once it begins to bloom, with its fresh, yellow flowers. You can also get a feel for how extensively it occurs through the tundra when its foliage turns bright red in the fall. On the west flanks of Agassiz Peak, I found it joined by moss campion, sticky polemonium and others in lovely little gardens (Fig. 4.18). A study on the Peaks found that Ross' avens acts as a nurse plant, facilitating the establishment of these other species, and then pays a price for its services by growing less (Michalet et al. 2011).

- Moss campion, ***Silene acaulis***, is common in fellfields and meadows in the tundra. In Arizona it occurs only on the Peaks; one of the closest moss campion populations is in the Sangre de Cristo Mountains in northern New Mexico—hundreds of miles away. As a true tundra plant, it pretty much stays above treeline; this, of course, makes it especially vulnerable to climate warming. It has a substantial range, described as a circumboreal species that may be represented by considerable variation, including subspecies and varieties. It occurs between the Peaks and the Arctic and then through Eurasia.

This is a classic cushion plant (Fig. 4.18, Plate 36); it has a puffier form than other spreading tundra species, and this protects it from the wind, which helps improve temperatures and moisture levels inside its little canopy. Moss campion assumes a cushiony form in harsher microsites, while growing flatter in milder sites. It has bright pink flowers and dark green, stiff, shiny foliage, which is reminiscent of moss. Even in its cushion form, it rarely exceeds several inches in height! It has been collected near the top of both Agassiz and Humphreys Peaks, where I'll bet it is in full cushion form. It can also be seen in the alpine meadow at the top of the Inner Basin. It is widespread in the Rockies. Hailing from the Rockies…somehow. There is a geneticist at the University of Arizona who intends to analyze the DNA of many of the isolated moss campion populations in the Southwest, to determine if they have diverged after becoming isolated from one another since the end of the Pleistocene. Such a study might also determine how long the various populations have been separated, and perhaps how many times they arrived.

-Parry's primrose, ***Primula parryi***, is one of the most beautiful plants that ever there was, and it lives above treeline (Fig. 4.18, Plate 40)! Unlike other tundra species, it gets up to a foot and a half tall in protected sites. It has large magenta colored flowers that have an unusual fragrance. It lives in lush settings, such as wet canyons in the spruce-fir zone, that go between there and the tundra. Though a high elevation plant, it attracts a large and diverse number of pollinators to its flowers (Miller et al. 1994). There are populations in the alpine meadow in Humphreys cirques, and in Snowslide Canyon near the spring. It has been seen as high as 12,000 ft (3,600 m) on the Peaks. Parry's primrose is an Intermountain West species, occurring from Arizona and New Mexico through Wyoming and into Utah and Nevada. It is thought to hail from ancestors in Asia and Europe. Through the Intermountain West, a number of other high elevation species have been derived from it. They are all endemic to the region, and it may be that their origins stem from the late Pleistocene (Kelso et al. 2009), showing again how the complex topography of the West and Pleistocene climates together led to considerable diversity.

- Sedges, ***Carex* spp**. There are five species of sedges in Arizona that occur only on the Peaks. Four of them occur at or above treeline, and include *Carex chaciolepis*, *C. elynoides*, *C. haydeniana* (Plate 34), and *C. albonigra*. Within the tundra, *C. albonigra* and *C. elynoides* are quite rare. Distributions like these are kind of mind-bending, really, reflecting species that need to live in the cold, and are thoroughly isolated by surrounding warmer conditions today. Most of these species are more common to the north, with the Peaks representing their southern edge. Sedges occur worldwide, though are most common in the cooler regions of both hemispheres. Like other high elevation species here, they are very vulnerable in the face of warming. There are at least 36 species of sedges associated with the Peaks. Sedges, like grasses, have come to rely on the wind for pollination, with many 'modern' adaptations that make the process more efficient. Their seeds may be dispersed by wind, water and by animals; apparently ants are fond of them.

- Annual plants are extremely rare in the tundra in most of the West. Presumably, conditions are just a little too unpredictable and harsh to support them over the long haul. In a way, *their* journey to the Peaks seems especially improbable. Several annual gentians, *Gentianella amarella* and *G. tenella* made it to the tundra, with *G. tenella*, or dwarf gentian (Plate 41), found in wetter alpine meadows, often in mats with other species, including moss campion and Ross' avens. They are beautiful little plants, with purple flowers and in Arizona they occur only on the Peaks.

-Plants with really wide elevational ranges: for one, sneezeweed (Plate 18) lives between 7,000 and about 12,000 ft on the Peaks (2,100 to 3,600 m): in my yard at 7,000 feet it will form a nice large rosette of leaves, but rarely blooms, unless I add extra water. It really finds its stride on Hart Prairie and blooms big time at about 8,500 ft (2,600 m), along with shrub cinquefoil (Plate 12), and while it makes it to treeline and above, it never blooms there, either. It spans a large range and reproduces where it can. Muttongrass (Plate 14) may take the prize, though, with a range spanning over 7,000 ft, from about 3,500 to 11,800 ft (1,000 to 3,600 m).

Mountain Shrubs

A great diversity of shrubs occurs on the Peaks. Three species of *Ribes*, or currants, occur across zones, with gooseberry currant making it all the way to treeline (Appendix I, Plate 9). Unfortunately, all of the *Ribes* species can vector white pine blister rust to both limber pine and southwest white pine, should it make its way to the Peaks (see mixed conifer section). Shrub cinquefoil, a low, shrubby species, occurs in the foothills of mountains through the West (Plate 12). On the Peaks it occurs on Hart Prairie, on the western foothills of the mountain. Mountain ash (Plate 21) and Rocky mountain maple (Plate 11) both favor moist areas in canyons and on the north side of the mountain. Bitter chokecherry fills a really important role in the Schultz Fire burn area these days: it has expanded its range through much of the burn site on steep slopes and, along with aspen, is helping to hold the little soil that survives on the slopes in place (Fig. 3.25). Of course, several species of elderberries (*Sambucus spp.*) occur through the forests, and produce abundant flowers and fruits that many animals depend on (Plate 10).

Willows

Three willows occur on the Peaks, and apparently mostly in the Inner Basin. Both *Salix bebbiana* and *S. scoulerana* are upland willows. There is a large population of *Salix bebbiana*—Bebb willow—on the western flanks of the Peaks on Hart Prairie (Plate 12). It is hanging on, but warming will not help it persist there, at such a low latitude. Bebb willow has an enormous distribution, including most of northern North America from Alaska south through the mountainous

Western Cordillera into northern Arizona and New Mexico. It may have migrated into the Peaks and surrounding areas during the Pleistocene, with its present shrinking populations speaking to a much wetter time. The largest willow stand on the Peaks—made up of Scouler willow—occurs in the upper Inner Basin at the base of Telemark chute, in the spruce-fir zone (Fig. 3.6). Willows don't have a huge presence on the mountain, but the scattered populations of these species speak to a time when they may have been more widespread, with more water.

Arroyo willow, *Salix lasiolepis*, occurs near the base of the Peaks along Schultz Creek, and south along the Rio de Flag. It anchors itself into these rocky drainages with massive underground root systems that more resemble tree trunks in width; this was determined by the excavation of the root system of a willow at the Museum of Northern Arizona. This species has many associated insects including gallformers, freefeeders including sawflies, leaf folding sawflies, cicadellids, grasshoppers, tortricids, and other butterflies, and parasitic wasps which parasitize most of them (Craig et al. 1988).

Mountain Mistletoes

There are four species of mistletoe on the Peaks, according to NAU plant ecologist Robert Mathiasen (Table 4.1). Southwestern dwarf mistletoe on ponderosa pine is the most common mistletoe on the mountain. In the mixed conifer zone, Douglas-fir dwarf mistletoe parasitizes Douglas-fir and occasionally corkbark fir. In the Inner Basin, in the spruce-fir zone, western spruce dwarf mistletoe parasitizes Engelmann spruce. And on Schultz Peak, bristlecone pine is being parasitized by bristlecone pine dwarf mistletoe. Jared Scott and Robert Mathiasen have described bristlecone pine dwarf mistletoe as endemic to the San Francisco Peaks and Kendrick Mountain (Scott & Mathiasen 2009). This endemic mistletoe has killed more than 90% of bristlecone pines on Schultz Peak (Fig. 4.16; Scott & Mathiasen 2009). Bristlecone's mistletoe occurs in 3 populations on the Peaks: on Schultz Peak, on the Weatherford Trail, and in the Inner Basin.

Mistletoe increases with tree densities… it's easier for their seeds to find suitable hosts. As parasites they can make forest trees more susceptible to drought and bark beetles, the combination of which can be lethal.

Mountain Cryptograms—Fungi, Lichens, Liverworts, Mosses and Ferns

There is little known about fungi, liverworts, mosses and lichens on the Peaks. To the rescue have come two ecologists in the area, Marylou Fairweather and Edward Smith, who contributed their mushroom list for the San Francisco Peaks to this book (Appendix II). Theresa Clark, University of Nevada-Las Vegas graduate student, and John Spence, NPS ecologist at Glen Canyon Recreational Recreation Area, contributed a list of mosses and lichens on the Peaks, and information about their habits (Table 4.3, Appendix I; Clark 2012, Spence et al. in prep.). According to Spence, more fieldwork is needed to understand the ecology and distribution of mosses and liverworts on the mountain; consequently, mosses will be added to the Peaks' plant list and to the total count of plant species, but not included in analyses of numbers of plants in the different

4.20. Honey mushroom, *Armillaria ostoyae*, is a pathogenic mushroom which damages many conifers on the Peaks. Photo courtesy of M. Fairweather & E. Smith

zones in this edition (Table 3.2, Appendix I).

With the cryptograms, we are definitely talking old! For some perspective, the earliest land plants evolved by the early Devonian from green algae, and included moss and liverwort-like forms. Cryptograms produce spores rather than seeds to reproduce. Seed production came much later, in the late Devonian! Liverworts, mosses and ferns are plants, while fungi are not, and lichens are kind of a mix of algae, fungi and bacteria. Being among the oldest living land plants, it's pretty cool that there are several species of liverworts living on the Peaks.

According to Clark, cryptograms weave the forests together, filling in the spaces between trees, shrubs, herbs and grasses as they stabilize soil and aid in the decomposition of downed wood and even rocks. Many cryptograms are more tolerant of both cold and dryness, and have much broader elevational ranges than more modern vascular plants. In fact, many lichens and mosses have the remarkable ability to dry out completely and persist in a dormant state for years. Upon being rehydrated, they can start photosynthesizing within seconds (Spence, personal communication 2018)!

Fungi on the Peaks

The Fairweather and Smith list of fungi is the most extensive list to date, with more than 100 species, including decomposers, parasites and mycorrhizae (Appendix II)! Recall that fungi, including molds, yeasts and mushrooms, break down a great deal of organic matter on Earth. They can also be highly mutualistic with plants. For instance, the mycorrhizae on the Peaks are crucial for the functioning of plants on the mountain, because they help their roots to draw in water and nitrogen. They include many of the mushrooms that we all love to harvest and savor, like red king bolete (Plate 45). The giant puffball, *Calvatia booniana*, is a saprophyte and breaks down organic matter (Plate 46). Meanwhile, Basquė garlic, *Pholiota squarrosa*, is a parasite or pathogenic type of fungus which can attack and damage plants and other fungi (Plate 47).

One species is particularly pathogenic in its relationship with conifers, and even aspen. Honey mushroom (*Armillaria ostoyce*) parasitizes the roots of Douglas-fir, Engelmann spruce and corkbark fir, ravaging their roots (Fig. 4.20, Appendix II). Younger conifers are most susceptible to this pathogen and can be killed by it. Even mature plants are more susceptible to it in drier climates. Treatment in forests sometimes includes clear cutting infested areas and replanting with more resistant conifers or hardwoods.

More than 50% of mushroom species occur in the mixed conifer zone, often in association with aspens, showing once again how fabulous this zone is. More species occur in ponderosa pine stands than in the spruce-fir zone (36 versus 25 species, respectively). Mushrooms 'bloom' or produce spores in spring and fall; species richness and abundance are greatest in fall, when warm temperatures and monsoon storms really come together (States & Gaud 1997). Fungal species' richness and abundance are also greatest where forests are dense. Not surprisingly, mushroom production is very limited on the Peaks during the winter. An exception is *Sclerogaster xerophila*, which can be found blooming near melting snowbanks (States & Gaud 1997).

Liverworts

In 1940, Elbert Little collected the first species of liverwort, *Lophozia longiflora* (previously called *L. porphyroleuca*), for the Peaks, found in its tundra meadows and spruce-fir forests (Appendix I; Little 1941). It is also known from the Rockies. Liverworts are very primitive plants, lack vascular systems, and produce spores rather than seeds for reproduction.

Lichens

Lichens are unlike land plants, as they consist of both a fungus and a green or blue green alga living together in a symbiotic fashion. Crustose, foliose and fruticose growth forms all occur on the Peaks (Table 4.3; Nash 2002). Little found lichens to be more common than mosses above treeline, particularly in exposed areas. Desiccation-tolerant crustose species are the most common and diverse, and their low stature helps reduce water loss. This will surely serve them well in the days ahead. Little (1941) noted that the most common lichen above treeline in the alpine fellfields was the dayglow-yellow crustose map lichen, *Rhizocarpon geographicum*. This hardy species is sensitive to air pollution and studied as a bioindicator in mountains around the world. Little also reported foliose lichens in the fellfields, including *Xanthoparmelia novomexicana*, the New Mexico rock shield lichen, and *X. conspersa*, the peppered rock shield lichen, which is recognized throughout the Peaks by its pale green lobes and brown appendages. Theresa Clark has collected foliose rock tripe lichens, *Umbillicaria* spp., on shaded boulders on

Table 4.3. Lichens of the San Francisco Peaks. Information courtesy of John Spence, Theresa Clark, Nash (2002), Little 1941, efloras.org.

Family	*Scientific Name*	Common Name	Ponderosa pine	Mixed conifer	Spruce-fir	Tundra
Rhizocarpaceae	*Rhizocarpon geographicum*	Map lichen				1
Lecanoraceae	*Rhizoplaca chrysoleuca* (historic synonym: *Lecanora rubina*)	Rock-posy lichen				1
Umbilicariaceae	*Umbilicaria* spp.	Rock tripe lichen		1	1	
Cetrariaceae	*Cetraria ereicetorum*					1
Parmeliaceae	*Brodoa intestiniformis*					1
	Xanthoparmelia conspersa	Peppered rock-shield lichen	1	1	1	
	Xanthoparmelia novomexicana	New Mexico rock-shield lichen				1
Usneaceae	*Usnea hirta*	Old man's beard lichen		1	1	

the Peaks. *Usnea hirta*, a fruticose lichen that lives on tree branches, is especially common on the Peaks (John Spence, personal communication 2018). It hangs from the trunks and branches of conifers in older stands in mixed conifer and spruce-fir forests (Plate 48). I have seen this species forever, and am so happy to learn what it is. There are nearly 2,000 species of lichens in Arizona alone!

Mosses

Theresa Clark and John Spence provided a list of 30 species of mosses for the Peaks (Appendix I). These little plants are mostly nonvascular—lacking vessels to transport water and nutrients. They form cushions or mats and periodically send up spore-producing capsules. The diversity and abundance of mosses should increase with elevation up into the spruce-fir zone, while harsh conditions in the tundra may limit their presence there (McCleary 1962). Despite occurring across a large elevation, most mosses on the Peaks spend most of the year dried and dormant, unless they occur near streams or springs.

Little (1941) described a white-haired moss, *Grimmia alpestris*, as common on dry, acidic rock outcrops above treeline on the Peaks, and Clark has observed this species in both the mixed conifer and spruce-fir zones. Clark expects rock-obligate species of *Grimmia* and *Orthotrichum* to be common and diverse on the Peaks. Little also found one of the most common mat-forming mosses, *Hypnum revolutum*, above treeline, mostly in crevices and shaded pockets in the fellfields, though it is also found in lower forests (Plate 49; Spence, personal communication 2018). *Ceratodon purpureus*, also called fire moss, is widespread on soil throughout the Peaks (Plate 50). It occurs in temperate to cold climates throughout the world! Other species include *Brachytheciastrum collinum*, which is common on logs in the mixed conifer zone, and *Pseudoleskeella tectorum*, a common mat-forming moss on rocks.

There are 24 species of ferns on the Peaks (Appendix I).

Mt. Elden plants…

Mt. Elden, a lava dome volcano just south of the Peaks (Fig. 2.1), is a very different place with regard to its plant communities. There are about 125 plant species that occur on Mt. Elden, but not on the Peaks. Some reach their northern limits on the warm south-facing slopes of Elden: Elden represents the end of the warm habitats for the desert crowd. Parry's agave (*Agave parryi*) reaches its northern limits here on the slopes of Mt. Elden and Kendrick Mountain, while its range extends as far south as Mexico City. There is a fern in Arizona, black spleenwort, *Asplenium adiantum-nigrum*, that occurs only on Mt. Elden (Appendix V)!

Animals On The Peaks Today

Mountain reptiles and amphibians

Not many reptile and amphibian species occur on the Peaks—but there are a few, and of course, some interesting stories. In 1851, purportedly at Leroux Spring, at the base of the western flank of the Peaks, Samuel W. Woodhouse discovered two amphibians, "one was a dark-brown toad, three inches long, with prominent cranial crests. The toad is marked with a highly definitive, whitish mid-dorsal stripe and irregular yellowish lines. [Herpetologist Edward] Hallowell named it *Bufo dorsalis,* in 1852, but thought it too similar to *B. lentiginosus* to publish it as a new species. In 1854, however, Charles F. Girard recognized it as unique and named it for Woodhouse, *Bufo woodhouseii* (Woodhouse's toad)" (Wallace & Hevly 2007). Today, Woodhouse's toad is known to be widespread in the southwestern and central United States, but it is not known from the Flagstaff area. So, perhaps it was here once and is now gone.

"Woodhouse also found salamanders abundant near the spring. Some were black above and marbled black and yellow below. Others were large and had yellow spots on the black upper body, yellow beneath. Hallowell was inclined to think they were of a new species and named them *Ambystoma nebulosum.* Today this salamander is considered a subspecies of the tiger salamander (*A. tigrinum* [*mavortium*] *nebulosum*)" (Wallace & Hevly 2007). So there's that: amphibians at a spring on the Peaks. Back then, Leroux Spring was surely perennial, a requirement for most amphibians, as well as being the headwaters for the mighty Rio de Flag. Tiger salamanders were found at tanks below Snowbowl Ski Area, and were also found at Veit Springs until the spring and its basin were modified, leaving little water to support amphibians.

The northern leopard frog, *Rana pipiens*, may have occurred on the Peaks at one time. Merriam stated that, "I have no doubt that this is the species which Hallowell records the name of *Rana areolata*, as collected by Dr. Woodhouse in the San Francisco Mountain" (Merriam 1890). This species is protected in Arizona, as it is disappearing through much of its historic range (Brennan & Holycross 2009). There is hope that releasing additional flow from Leroux Spring back into its drainage may create habitat suitable for this toad (see Springs below).

The greater short-horned lizard, *Phrynosoma hernandesi*, is known from ponderosa pine and mixed conifer forests up into the spruce-fir zone to 10,000 ft (3,050 m), and there is also a record of it from tundra habitats in New Mexico (Fig. 4.21; Brennan & Holyross 2009). Merriam described it as common in ponderosa pine forests on the Peaks. It is the most widely-distributed lizard in North America, occurring from Canada to Mexico.

The western or terrestrial garter snake (*Thamnophis elegans vagrans*) occurs at lower elevations throughout the Peaks, especially near springs and stock tanks in Hart Prairie. It has been found on nearby Kendrick Mountain at about 9,000 ft (2,750 m).

There is a report of a gopher snake near treeline on the San Francisco Peaks, according to *Reptiles and Amphibians of Arizona* (Brennan & Holycross 2009)! It is reported from most habitats below treeline, becoming abundant at about 6,500 ft (2,000 m).

There are a number of other reptiles and

4.21. The greater short-horned lizard, *Phrynosoma hernandesi*, in a mixed conifer forest on the San Francisco Peaks amphibians that *might* occur on and around the Peaks, according to Erica Nowak, NAU herpetologist, including the Arizona black rattlesnake (Fig. 4.22; see below), the terrestrial garter snake, *Thamnophis elegans*, which has been found at about 9,000 ft (2,750 m) in nearby Kendrick Mountain, and the boreal chorus frog, *Pseudacris triseriata*, which occurs in Hart Prairie and may occur higher on the mountain, due to substances in its blood which protect its cells and tissues from damage in freezing temperatures. The Arizona tree frog, *Hyla wrightorum*, is known from Baderville, a parkland at the base of the Peaks, where it is said to sound like a duck as it calls in mating season. The plateau fence lizard, *Sceloporus tristichus*, also occurs on basalt outcrops in Fort Valley, the ornate tree lizard, *Urosaurus ornatus*, occurs on the edge of nearby Flagstaff and maybe on the Peaks, though not very high up; and the many-lined skink, *Plestiodon multivirgatus*, occurs in ponderosa pine stands around the Peaks. Finally, the beautiful red, black and white banded Sonoran mountain kingsnake, *Lampropeltis pyromelana*, lives on nearby Mount Elden. It occurs through most of the Mogollon Rim as well as on many of the sky islands of the Southwest (Brennan & Holycross 2009).

The Arizona black rattlesnake, *Crotalus cerberus* (Fig. 4.22) may also occur on the Peaks. I am amazed that it even occurs around Flagstaff at 7,000 ft (2,100 m). It is black, perfect for blending with the ambient lava *and* for absorbing maximal heat. Occurring only along the Mogollon Rim, as many plants and animals do, this snake is nearly endemic to Arizona, except where the Mogollon Rim extends into New Mexico (Fig. 3.10, Plate 3).

4.22. Arizona black rattlesnake, *Crotalus cerberus*, near the Peaks. Courtesy of Gary Botello.

Mountain birds

This mountain is alive with birds. Most information on this bird story comes from four sources: research by C. Hart Merriam (1890), John Coons (1984), and John Haldeman (1973), interviews with Steven Carothers, Sandy Williams and Chuck Larue, and Ebird.org. Input from these different efforts yielded 161 species (Fig. 4.23, Appendix III). Ebird.org, an online site that lists birds encountered all over the world, includes sightings of birds by avid birders over a large elevational gradient on the Peaks. Ebird's lists of birds for the Peaks came from Hart Prairie, Snowbowl Ski Area, Humphreys saddle and the Inner Basin, and more than 150 species were recorded (EBird.org). Most species were found at Hart Prairie, Snowbowl and in the Inner Basin, while only 6 species were found along the summit trail, above 12,000 ft (3,600 m).

In the early 1980s, world-class birder John Coons studied birds through the seasons and along an elevational gradient between 8,000 to 11,000 ft (2,400 to 3,350 m) on the western flank of the San Francisco Peaks, (Coons 1984). He counted 40 species of birds, 29 of which occurred in the summer

4.23. Birds of the Peaks, clockwise from upper right: Northern Goshawk, *Accipiter atricapillus*; American Pipit, *Anthus rubescens*, Clark's Nutcracker, *Nucifraga columbiana*, and the Mexican Spotted Owl, *Strix occidentalis*. Courtesy of Gary Botello, except Goshawk (Paul Beier), Spotted Owl (Rick Ruess).

and fall (Appendix III). By July, summer residents declined at all elevations, probably due to more secretive parenting behavior and post-breeding dispersal. In fall, species numbers climbed again as southbound migrants passing through the region visited the mountain. The onset of winter had the most pronounced effect on numbers of birds on the mountain. Even some Northern Flickers, Brown Creepers, Steller's Jays, Pygmy Nuthatches, White-breasted Nuthatches and Clark's Nutcrackers retreated from high summer elevations to lower elevations in winter. By November, only 12 species, including the Golden-crowned Kinglet, were found up high. Further into winter, there were only 4 species—Clark's Nutcracker (Fig. 4.23), Mountain Chickadee, Red-breasted Nuthatch and Brown Creeper, found at 11,000 ft (3,350 m), through the heart of winter! These birds feed mainly on conifer seeds in winter, as insects are largely unavailable. The Mountain Chickadee was found at most elevations throughout the year, suggesting to Coons that this species is one of the true generalists on the mountain. Unlike other species, its numbers at higher elevations did not decline in winter!

By March, up to 15 species were counted in Coons' surveys, and all became more abundant by April, including Dark-eyed Juncos, Northern Flickers, Hairy Woodpeckers, American Robins, Yellow-rumped Warblers, and Ruby-crowned Kinglets. Many, including Hermit Thrushes, Yellow-rumped Warblers and Grace's Warblers, began establishing breeding territories then.

While birds were most numerous at lower elevations in May, numbers stayed low at high elevations, due to lingering snow. The number of birds at higher elevations in the mountain peaked as snow melted and more species visited higher elevations, and their communities were composed of one-third summer residents, one third permanent residents, and one-third Dark-eyed Juncos.

In the early 1970s, NAU ornithology student John Haldeman and colleagues studied breeding birds along an elevational gradient on the Peaks: in a ponderosa pine stand at the Fort Valley Experimental Station, and in a mixed conifer stand in Lockett Meadow in the Inner Basin, about 1,500 ft higher (460 m) in elevation (Fig. 2.6; Haldeman et al. 1973). They found 30 species nesting at the two sites: 23 species in the ponderosa pine forest, and 27 species in the mixed conifer stands, with hundreds of pairs of breeding birds at each site. There were more pairs of breeding birds in the mixed conifer stands than ponderosa pine stands. They also counted 18 species of birds that overwintered in the ponderosa pine forests on the Peaks, and 11 that overwintered in the Inner Basin...brrr. As with the study by Coons, Dark-eyed Juncos and Mountain Chickadees were most common among the high elevation winter birds (Haldeman et al. 1973).

Haldeman et al. counted a total of 69 bird species in the ponderosa pine forest, including nesting species. Violet-green Swallows, Pygmy Nuthatches, Gray-headed Juncos, Mountain Chickadees, and American Robins were the most common breeding species in the ponderosa pine forests. Plumbeous Vireos, Western Bluebirds, Northern Goshawks and Brewer's Blackbirds occurred only in the ponderosa pine forests, often associating with meadows amongst the forests.

In the mixed conifer forest in the Inner Basin, 41 species were seen, including 27 nesting species. The Warbling Vireo was the most common nesting bird there. Warbling Vireos, Red-naped Sapsuckers and William's Sapsuckers all nested in the extensive aspen stands. Five species—Warbling Vireo, Mountain Chickadee, Gray-headed Junco, Cordilleran Flycatcher and the Broad-tailed Hummingbird—comprised nearly 50% of the total nesting population. Warbling Vireos, Red-naped Sapsuckers, Western Tanagers, Clark's Nutcrackers, Williamson's Sapsuckers, House Wrens, Townsend's Solitaires and Cooper's Hawks occurred only in the mixed conifer site.

Just as the Dark-eyed Junco is widespread through the Peaks, it is also widespread and represented by 6 subspecies through North America. The Red-backed Junco is a resident of the Peaks and the Mogollon Rim across most of Arizona and New Mexico (Sibley 2003). The Gray-headed Junco, another subspecies, breeds farther north and then arrives in the area for the winter. It is very similar in appearance to the red-backed subspecies. We also have the Oregon Junco here, including in nearby Flagstaff; it has a dark grey hood.

Large predators, the Northern Goshawk and the endangered Mexican Spotted Owl, live in ponderosa pine and mixed conifer forests (Fig. 4.23).

The Northern Goshawk occurs throughout North America, but favors western and northern forests (Fig. 4.23). It was designated as a sensitive species in 1982, in USFS Region 3, which includes the San Francisco Peaks. The state of Arizona classifies it as a species of special concern. The U.S Fish and Wildlife Service turned down an application to list the goshawk as an endangered species due to a lack of information on its population structure. The Northern Goshawk is a very difficult species to study, making it all the more challenging to manage forests for its survival. Paul Beier, NAU ecologist, considers the goshawk to be doing pretty well on and around the Peaks.

It is found in many forested regions in the Northern Hemisphere. Studies have shown that the goshawk favors fairly dense ponderosa pine and mixed conifer forests with large trees on the Peaks and elsewhere through its circumpolar range, though it is also known from more open habitats (Hall 2001). It favors larger trees and dense forests for nesting. The goshawk often hunts from a perch, launching dramatic attacks when prey are spotted. Its large wings and long tail help it to navigate through such forests with speed and agility. Even paired males and females have separate, though overlapping, home areas during nonbreeding season. This sounds kind of nice to me. Males hunt for food, while females brood, defend and shelter young, maintain nests, and also hunt! Prey include small mammals like rabbits and squirrels, and large birds including jays, hawks, young turkeys, grouse and flickers, and small birds (Hall 2001). The Northern Goshawk responds fiercely to the presence of predators like Great Horned Owls and Mexican Spotted Owls. The goshawk most commonly nests in old trees, as well as dense stands of younger trees, and forests that have a structure which allows them to fly easily below forest canopies (Dickson et al. 2014). The goshawk favors nesting on north-northeast facing slopes, which are typically the coolest and most mesic, leading to productive understory growth, and consequently, more prey.

Northern Goshawks were studied in the vicinity of the Peaks in the early 1990s, and 34 of 41 known 'post-fledging family areas' were found to be occupied. Goshawks are known for high site fidelity, and tend to stay put.

Their populations fluctuate a lot between years, for a variety of reasons. Some reduced fledging success may occur during low precipitation La Niña years (Beier et al. 2007). Camping and logging constitute significant disturbances in their home ranges, and can cause them to leave their home area.

During the last glacial maximum, when extensive glaciers interrupted forests in the West, it is thought that the southern Colorado Plateau was an important refuge for goshawks (Bayard de Volo et al. 2013). Southwestern populations today are unique genetically, suggesting isolation in the Southwest during the last glacial maximum. Other forest birds on the Peaks, including Mountain Chickadees, the Dusky Grouse and Brown Creepers, are thought to have been similarly affected by glacial isolation, leading to population differentiation through their ranges (Bayard de Volo et al. 2013).

The Mexican Spotted Owl, *Strix occidentalis lucida*, was listed as an endangered species by the U.S. Fish & Wildlife Service in 1993 (Fig. 4.23). The move was unusual, as this owl is actually one of three spotted owl subspecies. The Mexican spotted owl has the largest distribution of the three, extending from Utah and the Colorado Rockies down into Mexico. The federal listing may have reflected the sense of urgency in the 1990s with the increasing loss of the spotted owl's habitat due to more timber harvests involving old growth trees, and inadequate regulatory means by which to protect it. This designation protected the spotted owls' territories or activity centers with buffers of 600 acres (240 ha) around nesting pairs. A more recent challenge for them comes from the occurrence of more large forest fires. Joseph Ganey, USFS Mexican Spotted Owl expert, and Carey Thompson, USFS biologist, report that a number of activity centers were lost on the Peaks during the Schultz Fire in 2010, and given their 'great fidelity' to nesting sites, such habitat loss can be especially significant. Thompson noted that following the 2010 Schultz Fire, the USFS observed two young Mexican Spotted Owls in an unburned pine tree, surrounded by burned forest. Can't you just imagine that conversation?

The Mexican Spotted Owl lives in old growth forests and canyons. It lives in big trees with diameters exceeding 24 inches (61 cm) in mixed conifer forests on the Peaks. It also occurs in nearby Mt. Elden and Dry Lake Hills. It nests in cavities and on platforms or witches' brooms caused by mistletoe infestations. Some migrate during winter, even as close as the Verde Valley, and some stay around. They lay eggs in early April and fledge young by June. These owls typically hunt at night, and may respond more to sound than visual action to detect prey. They are opportunistic, feeding on small mammals, as well as other birds and insects. They, in turn, may succumb to predation by the Northern Goshawk and the Great Horned Owl. While banded animals have lived to 18 years, it is thought that the Mexican Spotted Owl more often lives to 8 to 10 years of age. They do not hibernate.

Up in the mixed conifer zone, Clark's Nutcracker prefers southwestern white pine cones over limber pine cones when it attacks unopened cones, perhaps because they are less resiny (see mammal section below, Fig. 4.23). Way to go, limber pine. This changes, though, when cones open; then the nutcrackers also forage heartily on limber pine seeds. Cones on individual limber pine trees tend to open synchronously, which makes for efficient foraging by the nutcrackers. They get busy harvesting seeds as cones open into October. Earlier in the year the nutcracker harvests seeds in pinyon-juniper stands to the east of the Peaks. Its social caching habit enables this bird to breed earlier than most other birds in North America. Once cones open, many mammals, insects and birds including Steller's Jays, Williamson's Sapsuckers, Mountain Chickadees, and Red-breasted Nuthatches feed on their seeds.

John Coons suggests that both Red-breasted Nuthatches and Golden-crowned Kinglets breed in mixed conifer forests and spruce-fir forests. He has seen White-crowned Sparrows in the high elevation cirques in the western Inner Basin, which are near treeline. The White-crowned Sparrow population on the Peaks is thought to be its only breeding population in Arizona!

Above treeline, at the saddle, above 12,000 feet (3,650 m), Rock Wrens, Pine Siskins, Clark's Nutcrackers, White-crowned Sparrows and the American Pipit tough it out, mostly in the warmer months. The American Pipit, *Anthus rubescens*, breeds above treeline on the Peaks' tundra on the slopes of Agassiz in the upper Inner Basin; it is a tundra breeder (Fig. 4.23). It breeds in Arizona only on the Peaks. This species also breeds in tundra in the Arctic and in alpine tundra in the Rockies.

According to John Coons, some birds are more common today on the Peaks, perhaps due to warming. The Olive Warbler was rare in the 1980s, and today is regularly seen in the Hart Prairie area and in Lockett Meadow, while seemingly becoming less common in southern Arizona (John Coons, personal communication 2017). The Olive Warbler is in its own family, Peucedramidae, which is the only bird family that is endemic to North America. It also occurs in Mexico. It breeds on the Peaks and then heads south for the winter. Similarly, the Red-faced Warbler is more common on the Peaks and northern Arizona than it was even in the 1980s. The Three-toed Woodpecker is more commonly seen now following the bigger fires on the Peaks and elsewhere in the West (Coons, personal communication 2017).

Peregrine Falcons have increased through much of their enormous range since the 1980s, and have been observed migrating near the Peaks above treeline and in the vicinity of Little Spring (Chuck LaRue, personal communication 2017). This raptor is very widespread, with its populations having rebounded during the 1980s and 1990s, following the end of DDT use in the United States. It is known to breed in Oak Creek Canyon. Chuck LaRue, local Flagstaff birder, suggested that the peregrine may breed in the rockier outcrops higher on the Peaks; no one has ever looked, though he said he would!

The Dusky Flycatcher was described from Hart Prairie in 1939 by the famous ornithologist Allen Phillips (1939). Its type specimen was collected there, meaning that Hart Prairie specimens provided the basis for the description and name of this species! The Peaks and the Mogollon Rim are at the southern edge of its range.

Rare visitors to the Peaks include the rosy finches, the Black Rosy-Finch (*Leucosticte atrata*) and the Grey-crowned Rosy-Finch (*Leucosticte australis*), which have visited the Peaks and northern Arizona in winter! In February, 1967, both species were seen 'in numbers' at Snowbowl Lodge, where they were seen periodically until April (Carothers et al. 1973, 1978). The next record was for an individual found in 1968, on the slopes of Sunset Crater, about 6 miles (9.6 km) away. The latest records describe a small flock of them on the slopes of Sunset Crater in March 1971 and in 1997. They are regarded as irregular winter visitors on the Peaks. Local bird expert Chuck LaRue hadn't seen them in 20 years (personal communication 2017). They breed in the summits of many western mountains, and the Peaks lies at the southern edge of their winter range.

Mountain Mammals

The mammal list for this book is based on Merriam's list (1890), the list provided in Hoffmeister's book on mammals of Arizona (1986), and reviews and additions from NAU mammalogists Carol Chambers and Tad Theimer (Appendix IV, Fig. 4.24). There are 40 species of mammals that live on the Peaks for at least a portion of the year. Mountain sheep and grizzly bear were once reported for the Peaks, but are gone today.

Herbivores

Merriam's elk, *Cervus elaphus merriami* Nelson, was known only from the White Mountains, in eastern Arizona, and is now extinct (Hoffmeister 1986). Hoffmeister considered the type specimen of Merriam's elk to be quite large. It was actually Rocky Mountain elk, *C. elaphus nelsoni*, that was known from the portion of the Mogollon Rim that includes the Peaks. It was extirpated or eliminated locally from Arizona by 1910, probably by overhunting. Rocky mountain elk were introduced into

4.24. Some mammals of the Peaks: left to right, top to bottom: red squirrel, *Tamiasciurus hudsonicus*; yellow-haired porcupine, *Erethizon dorsatum*; mule deer, *Odocoileus hemionus*; mountain lion, *Felis concolor*; golden-mantled ground squirrel, *Callospermophilus lateralis*; Rocky Mountain elk, *Cervus elephas*; and black bear, *Ursus americanus*. Courtesy of Gary Botello, except black bear (Ellen Wade), and mountain lion (Donna Krucki).

the area several times, as initial introductions failed (Brown 2009). Elk on the Peaks today are the offspring of 54 animals introduced into the area around Bill Williams Mountain from Yellowstone National Park in 1928 (Fig. 4.24, Brown 2009).

Since their recent introduction, accompanied by a reduction of predators including wolves, mountain lions and bears, Rocky Mountain elk populations have exploded on the Peaks. Their impact on aspens especially is enormous and devastating (see Aspens above).

There used to be bighorn sheep on the Peaks. The best account comes from Merriam, recounting seeing 8-9 sheep scurrying up the ridgeline towards Humphreys, and then disappearing out of sight (Merriam 1890). They no longer occur on the Peaks.

The pronghorn, *Antilocapra americana*, was common around the Peaks when Merriam described it in 1889, occurring within a mile of his camp at Little Spring; it ranged between the Little Colorado River and the ponderosa pine forests around the Peaks (Brown 2009, Merriam 1890). This animal is the fastest-running mammal in North America, and second in the world, following the cheetah. Pronghorn have declined greatly in northern Arizona, due to roads, fencing and coyote predation, but are seen in low numbers in open parklands in ponderosa pine and pinyon-juniper zones.

Rocky Mountain mule deer, *Odocoileus hemionus*, are also common on the Peaks, often roaming above treeline in summer, and then occurring at lower elevations in winter (Fig. 4.24; Jim Heffelfinger, AGFD biologist, personal communication, 2017). While occurring statewide, northern Arizona mule deer are larger and darker than those in the deserts. The mule deer prefers forbs and shrubs, though it will graze young, green grasses. As with so many species, the south side of the Peaks represents the northern extent of white tail deer.

Abert's squirrel was first described by S.W. Woodhouse, from specimens he collected in ponderosa pine stands on the Peaks while on the Sitgreaves expedition in 1851 (Fig. 3.17; Brown 2009). Hoffmeister (1986) recognized at least 3 subspecies of *Sciurus aberti*, including *S. a. aberti*, *S a. kaibabensis*, and *S. a. chuscensis*, all in forests separated by lower elevation deserts. A recent study showed that there may be even greater genetic diversity among a number of this species' populations, with the migrations of ponderosa pine during the Pleistocene perhaps being influential (Lamb et al. 1997). Remarkably, Abert's squirrel is nearly entirely reliant on ponderosa pine for food and habitat. I just saw one of their nests yesterday, a pile of red pine needles on branches high up the trunk—southeast-facing, as usual. In winter, they feed on the inside of the bark of young pine shoots, casting the clippings of stem wood on the ground. Their distribution is limited to the southern latitudes of ponderosa pine's range, because persistent snow farther north makes them totally dependent on this dubious resource, and their populations fail when forced to feed on it for too long (See a discussion about this in Waring 2011). In warmer months they feed on many of the mushrooms associated with ponderosa pine's roots, and then developing cones and seeds. They also feed on Gambel oak acorns, when that species is present.

Squirrels and birds

While the relationship between limber pine and southwestern white pine is still unfolding, a great study on the western slope of the Peaks measured the interactions among a mammal and a bird and the cones of these closely-related pines (Benkman et al. 1984). Their cones and seeds are built to be accessible and desirable to the master of *seed dispersal*, Clark's Nutcracker, *Nucifraga columbiana* (Fig. 4.23): their seeds are very large and their cones are devoid of spines or prickles that some pine cones have to deter herbivores (Fig. 4.2). The red squirrel, *Tamiasciurus hudsonicus* (Fig. 4.24), and Clark's Nutcracker feed on both species' seeds and they both cache them to feed on through the winter. With this resource, Clark's Nutcracker is one of the earliest breeding birds in North America. The red squirrel is a major *predator* of conifer seeds in North America, and occurs throughout limber pine's range, from the Peaks all the way to Canada. The red squirrel is the noisy one you hear in mixed conifer stands—scolding you with chtchtcht or such. In the Peaks' study, the squirrels protected developing cones from other seed predators with aggressive territorial behavior. The red squirrel consumed more that 80% of the cones on trees studied, clearly the greater consumer of the trees' resources. The squirrel harvests whole cones when they are still closed, which kills the developing seeds, precluding their establishment. It has to cache them before they open, as more mature cones might open in the cache, spoiling the seeds and making them inedible. They cache cones at the base of trees.

Developing limber pine cones were heavily covered with resin, requiring red squirrels to stop often for 'face-wiping' while harvesting cones. Because of this, it took them longer to harvest and cache unopened cones from limber pine than southwest white pine...pretty good trick, limber pine (Benkman et al. 1984). The plants don't want their developing seeds to be disturbed. It has been shown that the production of resin is expensive energetically, making it a big deal. The recent hybridization studies show that both pine species actually produce a fair amount of resin (Kristen Waring, personal communication, 2018), so this story is still unfolding!

By contrast, Clark's Nutcracker harvested ripened seeds from opened cones, with resin largely dried and fallen away, making for easier and faster seed collection. The nutcracker is uniquely adapted for collecting such big seeds, with its sublingual pouch enabling it to carry many seeds for caching (see Waring 2011). By caching mature seeds, the nutcracker is a principal disperser and cultivator of these big seeds, as some are lost track of and germinate instead of being eaten. I mean, lets face it, by caching ripe seeds in soil, they are setting them up to germinate. It's that simple and fabulous. The nutcracker is a social bird, and collective harvesting and caching can produce quite a winter resource. The nutcracker harvested up to 70% of cones in forests where the squirrel was absent. Where the squirrel was present, the nutcracker harvested less than 10% of the trees' seeds. Ironically then, it is the nutcracker that actually disperses the pines' seeds and helps them to establish. So, the one that takes the least contributes the most to the conifer's fitness.

Once limber pine and southwestern white pine cones open, a kind of free-for-all ensues, with Clarks' Nutcracker followed by other seed-eating birds including Steller's Jays, Mountain Chickadees, and Red-breasted Nuthatches, and cone insects that also forage on the seeds. DePinte (2016)

4.25. Red squirrel cone-shredding pile beneath a Douglas-fir.

found that up to 80% of cones were infested with cone insects. The golden-mantled ground squirrel and the gray-collared chipmunk gather seeds from cones on the ground, and there is even a report of a black bear excavating red squirrel cone caches and eating them (see below).

The red squirrel also goes to town on Douglas-fir cones and other pine cones in search of their seeds, shredding them down to a stub, and leaving deep mounds of scales piled up feet deep at the base of trees (Fig. 4.25).

Porcupines occur up to treeline on the Peaks (Fig. 4.24). While Merriam never observed them in 1889, he found ample evidence of them, on seeing various conifers with gnawed bark, with bristlecone pine at treeline being a favorite (Merriam 1890, Brown 2009). Porcupines were seen on the saddle between Humphreys and Agassiz peaks, and a study in the 1920s found that their dens occurred high up on the Peaks, mainly on south-facing slopes (Taylor 1935). In summer, these creatures favor herbaceous plants and all kinds of seeds, and in winter they rely more on cones and needles and even the inner bark of conifers. In some areas around the Peaks, porcupine damage to mid-aged ponderosa pine exceeded 70% of trees (Taylor 1935). Coconino National Forest silviculturalist, Andy Stevensen, recalled a friend's dog being attacked by a porcupine above treeline on Humphreys saddle several years ago.

Merriam did report a weasel near the summit of the Peaks, but never encountered one (Brown 2009). Tad Theimer, NAU mammalogist, suggests that this was unlikely. Given that they are more visible/evident in other parts of their North American range, they are considered much less abundant in Arizona, even in appropriate habitat.

Predators

Not Merriam, Woodhouse, Kennerly, Mollhausen, Coues or the naturalists on the Wheeler survey mentioned seeing wolves on the Peaks in the 1800s (Brown 2009).

There were grizzly bears (*Ursus horribilis*) in northern Arizona until the early 1900s. Merriam cites the accounts of Woodhouse and Kennerly, on the Sitgreaves expedition in 1851, that grizzlies were abundant on the Peaks; he personally never saw them. This is the only account of their presence on the mountain that I know of. There is a great story, though, of the trapper and explorer Antoine Leroux being impressed with evidence of very high numbers of grizzlies near Sitgreaves Mountain, just west of the Peaks (Merriam 1890): it was wintertime and there was an abundance of bear tracks in the snow, and scratches on a pond that had frozen over, sending the troupe farther south.

According to Hoffmeister (1986), grizzlies may have been abundant along the Mogollon Rim, while persisting longest in the White Mountains, where the last known grizzly in Arizona was killed on Baldy Peak in 1925. They were fiercely exterminated by settlers. Today, there are considered to be two subspecies of brown bears in North America. Being the consummate 'splitter' that he was, C. Hart Merriam described nearly 80 subspecies of grizzlies, based on small levels of variation that today are recognized as naturally-occurring within species (Merriam 1890).

The grizzly or brown bear reaches 1,000 pounds or more in weight. While it can make short order of prey, it is also very fond of plant roots and bulbs, which its long claws enable it to dig for.

There are black bears on the Peaks today (Fig. 4.24). They are generally shy and run away from encounters. I do know someone, though, who was awakened from a nap on the edge of a lovely meadow, by the sounds of a bear attacking a baby elk! A recent study found that black bears harvested southwestern white pine seeds on the north side of the mountain, and in one instance, even unearthed and ate a red squirrel's seed cache (Mattson & Arundel 2013).

Mountain lions are well-known from the Peaks and surrounding low mountains including Dry Lake Hills and Mt. Elden, where their huge paw prints are often found in soft soil (Fig. 4.24). A study by the USGS (Mattson 2007) followed 10 cats that had been collared and tracked for several years, and found that many had enormous ranges on the Peaks and along the Mogollon Rim (Fig. 3.10, Plate 3). One female had a range that encompassed the entire mountain! Elk and deer were their primary prey (>80%), with small mammals also being taken. Mountain lions travelled farther in winter than in warmer months in search of prey. Merriam's group killed an old female lion in the spruce-fir zone in late afternoon. Her stomach was full of small mammals, including a red squirrel, a mountain chipmunk, two pocket gophers and a mountain vole which she hunted during the day (Brown 2009). Seems harsh.

Bobcats are probably common. They were seen by Merriam while camping at Little Spring, when they came to drink water at night.

Mountain Insects

Some of the best records of insects on the Peaks come from people who study insects that attack its forest trees. According to Monica Gaylord, USFS, a number of forest trees on the Peaks experienced insect outbreaks in the 1950s and early 2000s, on the heels of significant regional drought, and large die offs occurred within many species. The greatest attacks have been by piñon ips on pinyon, *Ips* bark beetles attacking ponderosa pine, and multiple biotic and abiotic agents attacking aspen.

Most of the forest species are preyed upon by bark beetles and defoliating insects, which are best described in the sections of individual tree species above (see Zones above, Table 4.2). Drought and disturbance seem to drive most outbreaks. In addition to heavy mortality due to bark beetles, defoliators including western spruce budworm (*Choristonerua occidentalis*) attacked many conifer species on the Peaks. A western spruce budworm outbreak coincided with drought in the 1950s and early 2000s. Western spruce budworm defoliates Douglas-fir and white fir, and to a lesser degree Engelmann spruce, blue spruce and corkbark fir (Lynch et al. 2008).

About half way up Snowbowl Road, you can see an emerging outbreak of forest insects and dwarf mistletoe on the ponderosa pine in the mixed conifer forest there. Many large pines have canopies dotted with dead apical twigs. Some trees are also infested with mistletoe and bark beetles, with some mortality resulting, according NAU forest ecologist Tom Kolb (personal communication, 2016).

Pollinating insects are getting a lot of attention on the Peaks these days. Ecologists at the Merriam-Powell Research Center at NAU report more than 350 species of bees on the mountain, along with 96 species of flies and 70 species of moths and butterflies; many of them are potential pollinators (McCabe et al. 2018). Several studies found that pollinating bees are more abundant and more diverse at lower elevations (Stevens et al. 2007, McCabe 2015). They found that bees drop out at higher elevations and flies take over as pollinators. Lindsie McCabe's study showed that bees and flies are most abundant and diverse in the mixed conifer zone, the sweet spot. McCabe found communities of bees and flies on Kendrick Mountain to be quite different from those on the Peaks, and regards the recent, extensive fires there to be influential.

While there may not be as many pollinators at higher elevations, they apparently do show up for gorgeous Parry's primrose in the tundra (Fig. 4.18, Plate 40); it is thought to attract numerous pollinators. A study of this species in Colorado found that it produces copious nectar and attracted large numbers of pollinators in 15 insect families, including lots of bees, 9 species of *Bombus* alone, butterflies, and flies, and that was at a site over 12,000 ft (3,650 m; Miller et al. 1994)!

There are few species of mosquitoes on the Peaks. In fact, there are few species of mosquitoes in northern Arizona, perhaps due to its aridity (Stevens et al. 2008). There is one species, though—*Ochleratus fitchii*–that occurs in mixed conifer forests and meadows of the Peaks, emerging from meltwater from snowbanks (Stevens et al. 2008)! The Mogollon Rim is the southern limit of this species' distribution.

There are a number of interesting invertebrates living above treeline on the Peaks. Many insects, including the

1. The distribution and age of major lava flows on the San Francisco Peaks. Note that the youngest andesite flow nearly encircles the mountain. After Holm (1986).

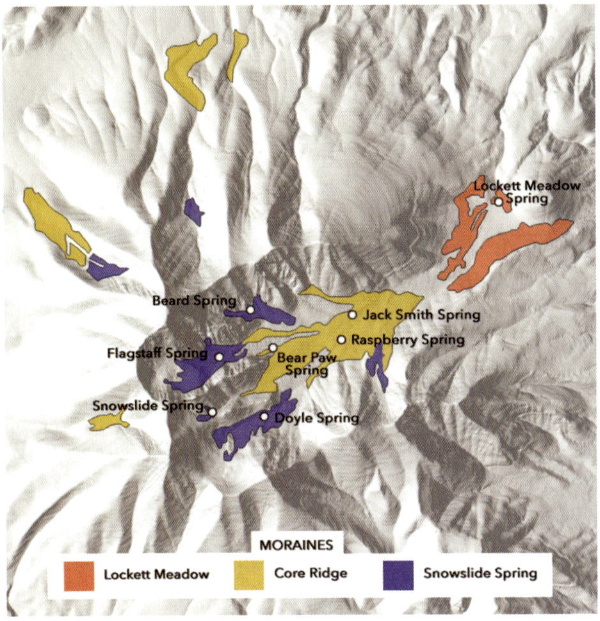

2. The extent of three glaciers in the Inner Basin. After Péwé & Updike (1976).

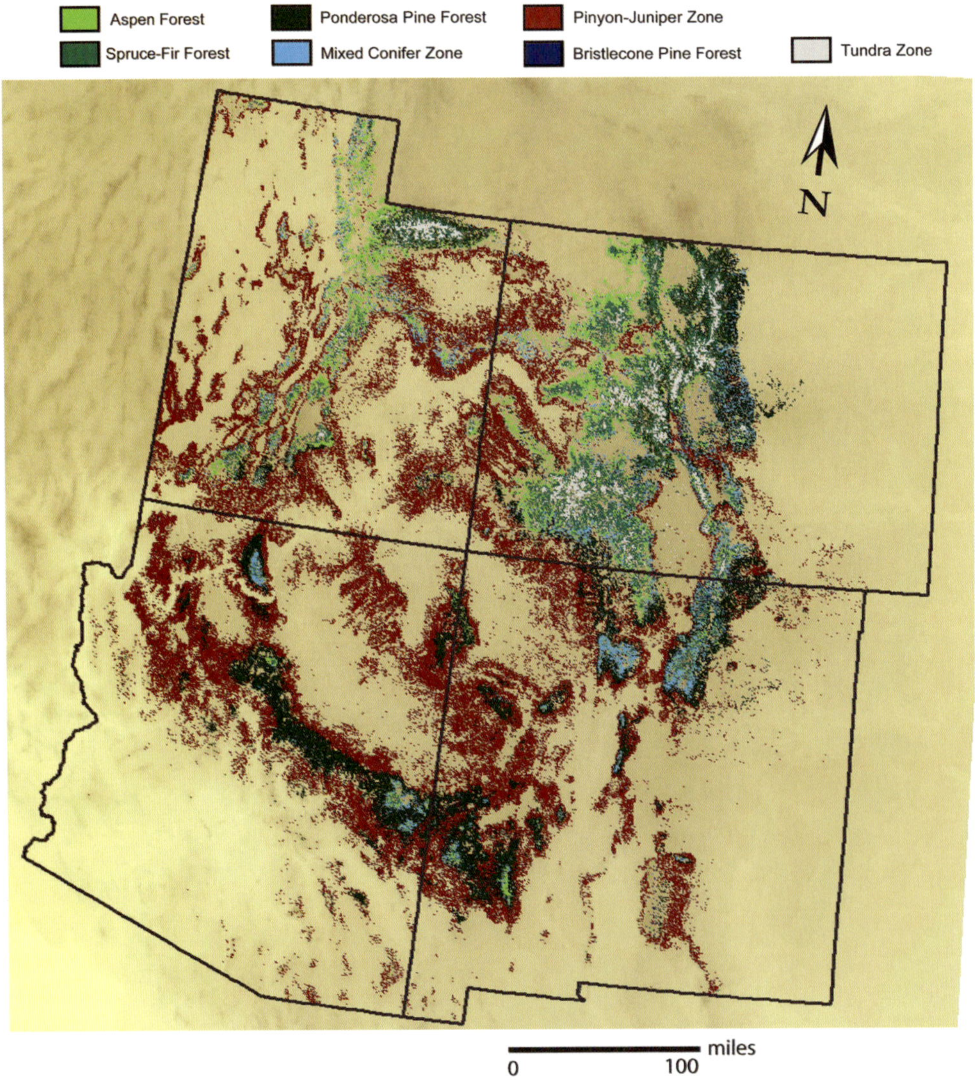

3. Colorado Plateau vegetation map. Courtesy of John Vogel, DOI geographer, and 2005 USGS GAP Data. Note the presence of the hefty Mogollon Rim, as the southern edge of the Colorado Plateau in Arizona. It traverses the state diagonally, dipping down into New Mexico at its eastern end.

4. San Francisco Peaks vegetation map, by plant zones. Courtesy of John Vogel, DOI geographer, and USGS GAP Data.

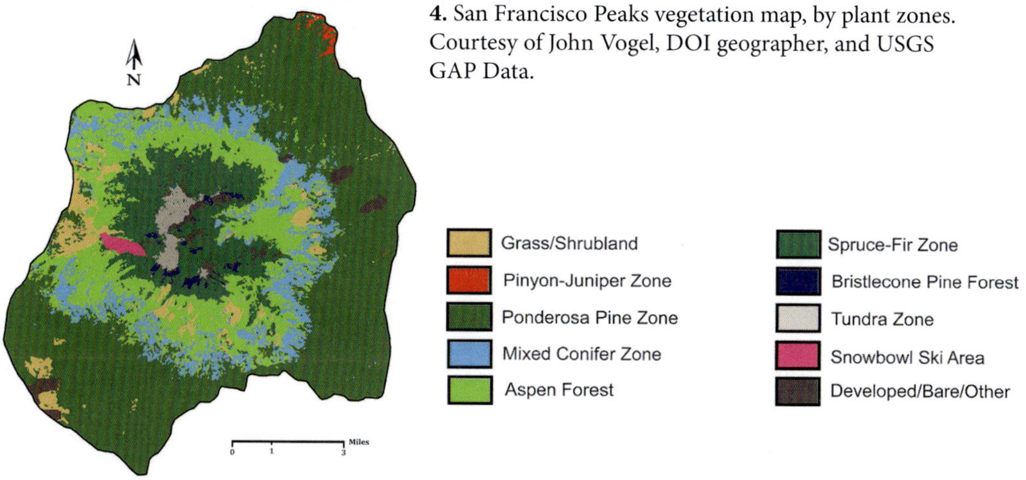

5. Precipitation and evaporation across the Coconino Plateau. Courtesy of Bills et al. (2007).

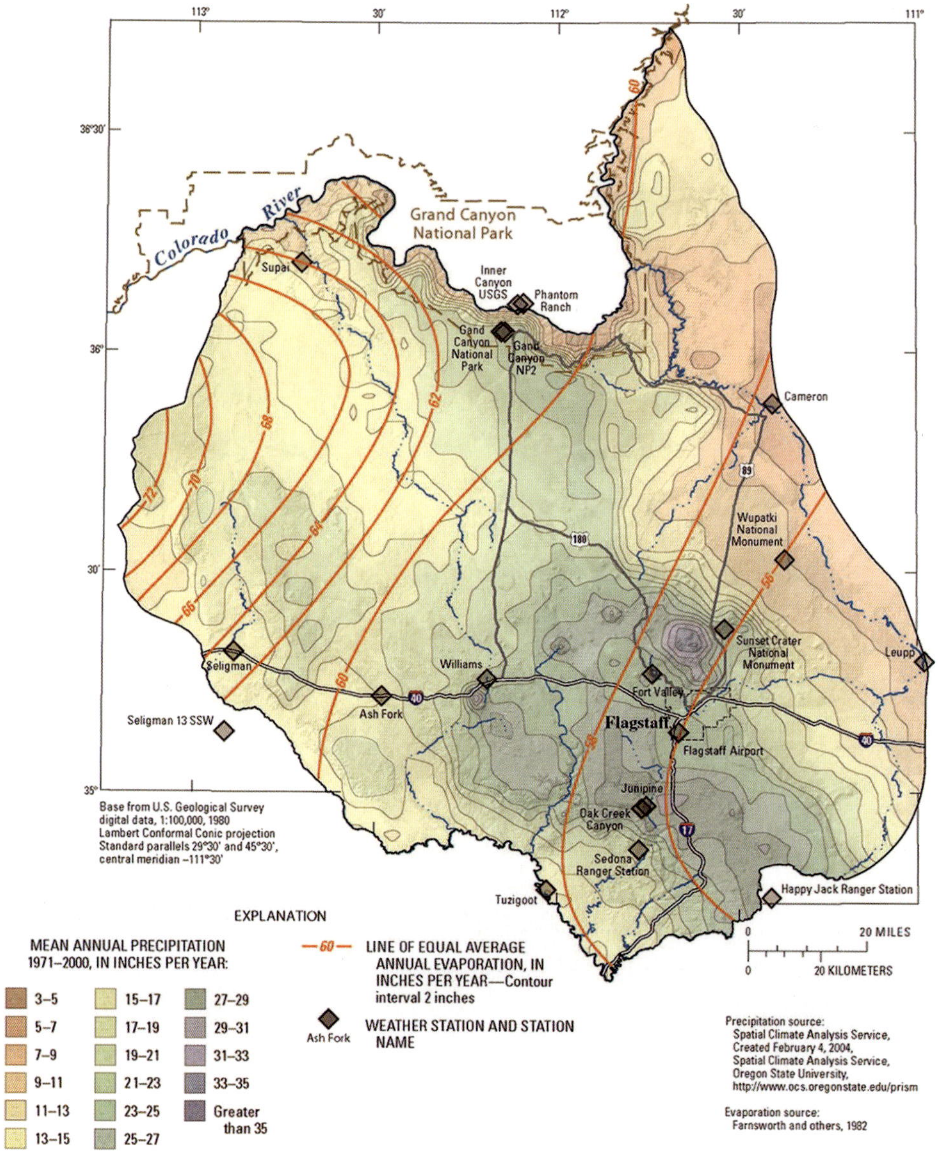

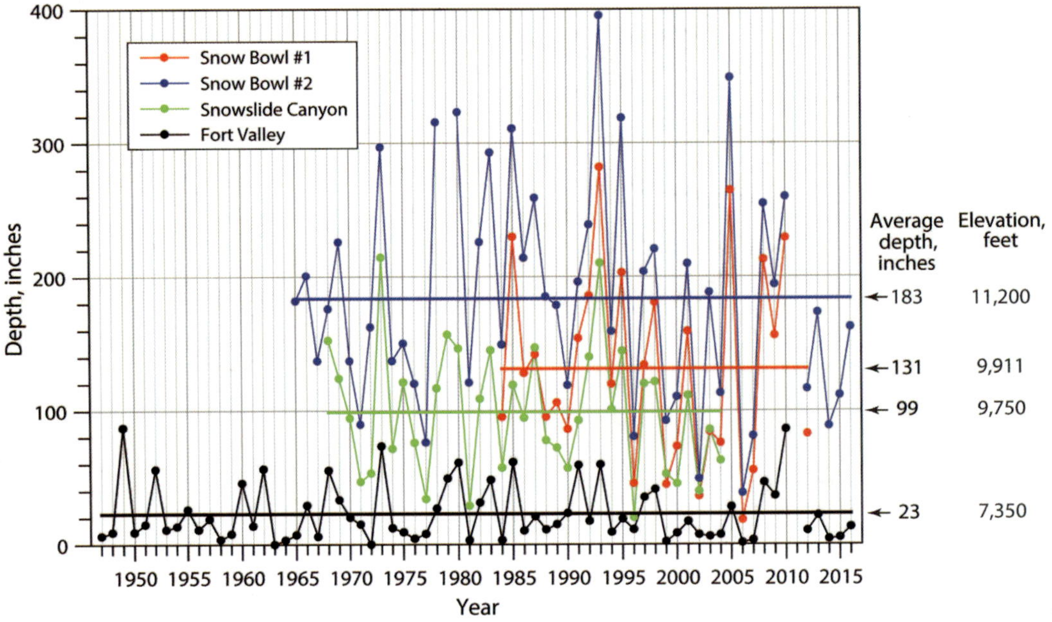

6. Annual snowpack across elevations on the Peaks, based on snow course data. Courtesy of Richard Hereford.

7. Wax currant, *Ribes cereum*, occurs at lower elevations on the Peaks. Its smooth red berries are an important food source in the forest.

8. Orange gooseberry, *Ribes pinetorum*, occurs in the Peaks' upper forests. Its fruits and branches are armored with spines to discourage herbivores. Surprising for a fruit-bearing plant. Courtesy of Max Licher.

9. Gooseberry currant, *Ribes montigenum*, occurs through all zones!

10. Red elderberry, *Sambucus racemosa*, produces copious berries in the mixed conifer and spruce-fir zones

11. Rocky Mountain maple, *Acer glabrum*, graces wetter canyons and shaded slopes. Big trees grow in the north-facing canyons on the road to the Inner Basin. Courtesy of Max Licher.

12. Shrub cinquefoil, *Dasiphora floribunda*, is especially happy on the slopes of Hart Prairie, seen here with Bebb willow. Courtesy of Max Licher.

13. American vetch, *Vicia americana*, is a nitrogen-fixing legume common to the Peaks' forests. Courtesy of Max Licher.

14. Muttongrass, *Poa fendleriana*, occurs through all elevations on the Peaks.

15. Arizona fescue, *Festuca arizonica*, graces high meadows in the four-corner states. Courtesy of Max Licher.

16. Mountain muhly, *Muhlenbergia montana*, is a high meadow grass in the four-corner states. Courtesy of Max Licher.

17. Pleated gentian, *Gentian affinis*, grows with Arizona fescue on Hart Prairie.

18. Sneezeweed, *Hymenoxys hoopsii*, occurs from Hart Prairie to treeline. Courtesy of Max Licher.

19. Parry's bellflower, *Campanula parryi*, occurs in forest meadows on the Peaks. Courtesy of Max Licher.

20. Silver lupine, *Lupinus argenteus*, another nitrogen-fixing legume common to the Peaks' forests. Courtesy of Max Licher.

21. Arizona mountain-ash, *Sorbus dumosa*, a rose, occurs in the Peaks upper forests, and produces white flowers and then red fruits. Courtesy of Max Licher.

22. Western brackenfern, *Pteridium aquilinum*, understory in an aspen stand on upper Hart Prairie.

23. Richardson's geranium, *Geranium richardsonii*, occurs in the understory of all the Peaks forests. Courtesy of Max Licher.

24. Greenflowered wintergreen, *Pyrola chlorantha*, occurs in spruce-fir forests on the Peaks.

25. Great red paintbrush, *Castilleja miniata*, a partial plant parasite, occurs in meadows and forest margins. Courtesy of Max Licher.

26. Canadian white violet, *Viola canadensis*, occurs through the Peaks' forests to treeline. Courtesy of Max Licher.

27. Golden columbine, *Aquilegia chrysantha*, occurs all the way to the tundra on the Peaks. Courtesy of Max Licher.

28. Beautiful fleabane, *Erigeron formosissimus*, occurs in open meadows and forest understory on the Peaks. Courtesy of Max Licher.

29. Little false Solomon's seal, *Maianthemum stellatum*, is a staple of forest understory communities on the Peaks. Courtesy of Max Licher.

30. Pygmyflower rockjasmine, *Androsace septentrionalis*, occurs from ponderosa pine forests to treeline. Courtesy of Max Licher.

31. Fairy slipper, *Calypso bulbosa*, one of the Peaks' five orchids occurs in shady canyons and along the edges of subalpine meadows. Courtesy of Steve Garro.

32. Twinberry honeysuckle, *Lonicera involucrata*, in the Inner Basin is heavily browsed by elk. Courtesy of Max Licher.

33. Whipple's penstemon, *Penstemon whippleanus*, is part of the tundra community. Courtesy of Max Licher.

34. A decent population of cloud sedge, *Carex haydeniana*, below Humphreys saddle; it occurs only in the Peaks in Arizona. Courtesy of Max Licher.

35. Ross' avens, *Geum rossii*, above treeline with many pollinators on the western slope of Agassiz.

36. Moss campion, *Silene acaulis*, above treeline. Courtesy of Max Licher.

37. Sticky polemonium, *Polemonium viscosum*, with Ross' avens above treeline on Agassiz Peak.

38. San Francisco Peaks ragwort, *Packera franciscana*, Peaks endemic, above treeline. Courtesy of Max Licher.

39. Graceful buttercup, *Ranunculus inamoenus var. subaffinis*, a high elevation Peaks' endemic. Courtesy of Max Licher.

40. Parry's primrose, *Primula parryi*, pouring down Snowslide Canyon.

41. Annual gentian, *Gentianella tenella*, occurs above treeline on the Peaks and other western mountains. Courtesy of Max Licher

42. The gorgeous, perennial gentian, *Gentianopsis barbellata*, occurs only on the Peaks in Arizona, with the nearest related population in northern New Mexico. Courtesy of Max Licher.

43. Spiked woodrush, *Luzula spicata*, Courtesy of Max Licher.

44. Alpine mountain-sorrel, *Oxyria digyna*, above treeline; it occurs only on the Peaks in Arizona. Courtesy of Max Licher.

45. Red king bolete, *Boletus edulis*, a mycorrhizal and tasty fungus, occurs through the Peaks' forests.Courtesy of Marylou Fairweather and Edward Smith.

46. The giant puffball, *Calvatia booniana*, is a saprophyte in ponderosa pine forests, here with Campanula. Courtesy of Marylou Fairweather and Edward Bennett.

47. Basque garlic, *Pholiota squarrosa*, is a parasite in aspen forests. Courtesy of Marylou Fairweather and Edward Bennett.

48. Old man's beard, *Usnea hirta*, graces the branches of trees in older mixed conifer and spruce-fir stands. Courtesy of Theresa Clark.

49. A common matt-forming moss, *Hypnum*, occurs above treeline and in spruce-fir forests. Courtesy of Theresa Clark.

50. Fire moss, *Ceratodon purpureus*, is widespread on soil through the Peaks, and in cold and temperate climates throughout the world! Courtesy of Theresa Clark.

4.26. High elevation butterflies on the Peaks: Left, Alberta Arctic butterfly, *Oeneis alberta*; Right, Arizona silvery blue butterfly, *Glauchopsyche lygdamus arizonensis*. The Alberta Arctic butterfly is orange with dark brown spots on its wings, while the Arizona silvery blue is silvery blue like most lyceanids. Courtesy of Tim Cheknis.

hornworm wasp, *Tremax columbia*, ladybugs, *Hippodamia convergens*, and big-bodied syrphid flies, 'hilltop' at the top, that is, they congregate at the top of Humphreys, probably for mating. There are also springtails, ground beetles, perhaps a new species of centipede, and wasps, which live on rocky slopes. There is even a land snail that lives above treeline, collected by Harold Colton in the 1930s, and it still resides there today (Larry Stevens, personal communication 2017)! While there are a number of dragonflies on the Peaks, only species such as darners make it to the highest elevations. Several species of tiger beetles also make it to near treeline (Stevens & Huber 2004).

The Alberta Arctic butterfly, *Oeneis alberta*, and the Arizona silvery blue butterfly, *Glochepsyche lygetamas*, are examples of insects that are abundant only on the Peaks, and a few other southwestern mountains, including the White Mountains. They are not found at lower elevations in the region, and may be isolated on their respective mountains (Larry Stevens, personal communication 2018). The Arctic butterfly is abundant in aspen stands and mixed conifer meadows on the Peaks (Fig. 4.26).

There are reports of tarantulas as high as 10,000 feet (3,050 m) on the Peaks. Larry Stevens suggests that a southwestern tarantula, *Aphonopelma marxi*, is the most likely candidate.

Tundra insects are yet to be studied, though I have seen many flies on Ross' avens' yellow flowers above treeline! Given the large number of flowering plant species in the tundra, there may be an interesting array of insects associated with them.

Nonnative plants and animals

There are not many nonnative plants on the Peaks (52 species, Appendix I). I noticed on a trek around the Peaks that this is a pretty pristine mountain. Fisher & Fulé (2004) found that Kentucky bluegrass was the most commonly encountered nonnative grass on the Peaks. This species is interesting, it is so widely-distributed globally, and so diverse genetically, I don't think anyone really knows its origin. If you look at a map of its North American distribution, it is everywhere! The Snowbowl developers have seeded the ski slopes with several nonnative grass species, including smooth brome (*Bromus inermus*). Fisher & Fulé (2004) found more nonnative species at lower elevations than at higher elevations; they suggest that the forests serve as buffers against nonnative plants. Viva la buffers. There are places at lower elevations where cheatgrass is starting to rear its ugly head. Fisher and Fulé suggest that cheatgrass doesn't fare well in shade or dense litter. Of course, dandelions and Dalmation toadflax occur everywhere. Larry Stevens makes the case for the humble dandelion, with its early blooming in the region helping pollinating insects through precarious climates.

Rocky Mountain elk numbers are way too high to sustain aspen populations. The spruce aphid, which was first found here in 1999, is a serious pest of both spruce species.

— SECTION 5 —

Climate, Water, and the Pull of the Mountain and the Mogollon Rim

Climate

In describing the climate of the Peaks more than 100 years ago, C. Hart Merriam (1890) first described climate in the tropics. "The tropics are characterized by great uniformity of temperature, the daily and yearly fluctuations being insignificant. The absence of a marked fall in temperature at night is due in great part to the large quantity of moisture in the atmosphere. This moisture acts in two ways: 1. by diminishing the loss of heat by radiation; 2. by directly increasing the temperature of the atmosphere. As stated by [Alfred] Wallace, "the warmth given off by the heated Earth is very largely absorbed by it [the aqueous vapor], thus raising the temperature of the air; and as it is the lower strata of air which contain most vapor, these act as a blanket to the Earth, preventing it from losing heat at night by radiation into space."

In great contrast, "The excessively dry atmosphere of Arizona acts in exactly the opposite way, interposing no obstacle to free radiation and presenting no medium to retain the heat given off at night. Hence, the change in temperature from day to night is always great. The summer heat of Arizona, except on the high mountains, is greatly in excess of the summer heat of the tropics, while the winter temperature is vastly lower…" (Merriam 1890).

Several things set up the area around the Peaks to be rather dry. The overall weather is dominated by the Hadley Cell, through which moist, warm tropical air rises, loses its moisture, migrates north and descends as dry air in the Southwest. The result is a high pressure system with little precipitation (Vankat 2013). Throughout much of the year, the mountain is marked by frequent periods of dry weather and pervasive sunshine, interspersed with storms. Of course, the Sierra Nevada of California exert a rain shadow effect which also helped to create all of the southwestern deserts of North America (Warner 2004).

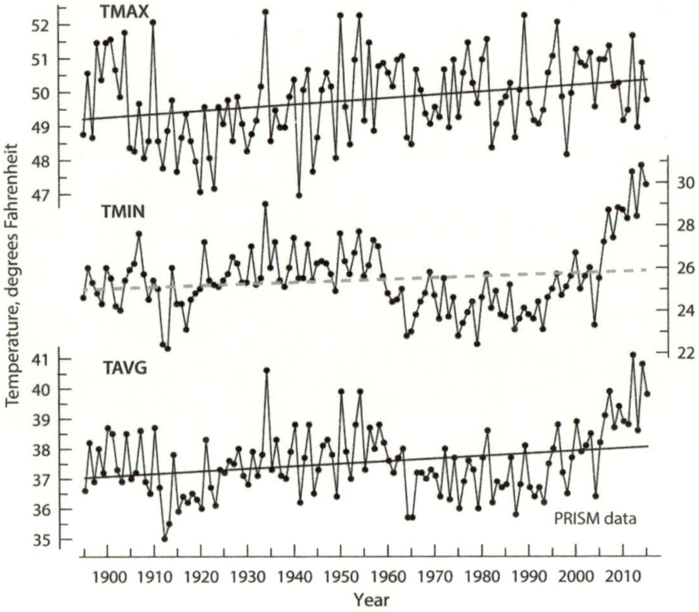

Figure 5.1. Long-term temperature estimates for Humphreys Peak: a time series of annual minimum (TMIN), maximum (TMAX) and average (TAVG) annual temperatures, 1895-2015. Long-term maximum and average temperatures have increased significantly (P< 0.05) since 1895 at 2.4 and 0.9 Fahrenheit degrees per century, respectively. Average minimum temperatures have not increased significantly, although the minimum temperature has increased since the mid-1980s to the highest levels since 1895. Graphic courtesy of Richard Hereford, USGS. Data were obtained March 24, 2017 from PRISM Climate Group, Oregon State University, http://prism.oregonstate.edu. A four square kilometer temperature grid was extrapolated to Humphreys Peak.

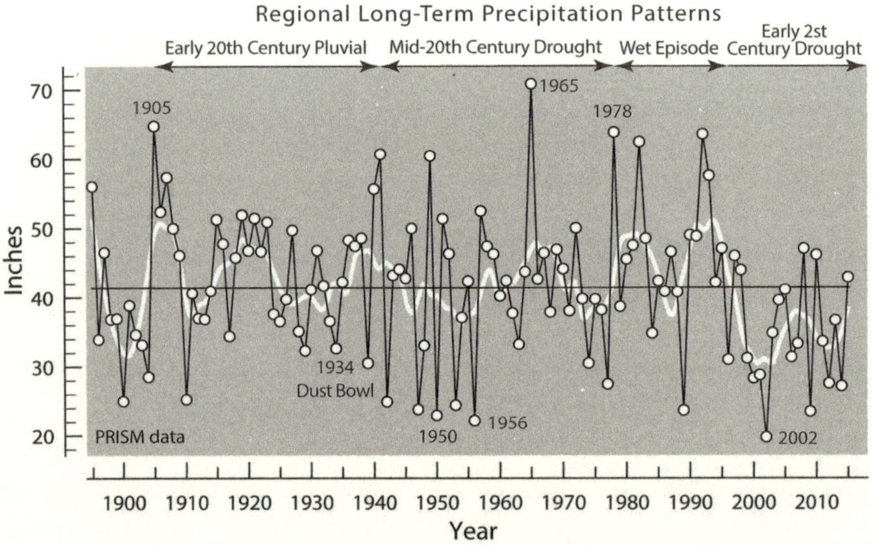

5.2. One hundred and twenty-one years of annual precipitation in the Humphreys Peaks-Inner Basin area show the main precipitation episodes identified in the Southwest. Black line is long-term average annual precipitation. White line is the smoothed annual precipitation showing multiyear wet and dry episodes. The three wettest and driest years are labeled. Graphic courtesy of Richard Hereford, USGS (data source: Prism Climate Group, http://prism.oregonstate.edu).

Some climate scientists predict that with global warming the Hadley Cell will expand, and because of that and other factors, the Southwest is moving into a climate state unlike any known in the 'instrumental' record. Oceanographer Richard Seager thinks that the droughts of the past, like the Dust Bowl of the 1930s and the 2000-2004 drought, will become the norm of the future (Seager et al. 2007). Weather data for the Peaks show both increasing temperatures and a decrease in precipitation in the 21st century (Fig. 5.1, 5.2; Hereford 2014).

Water Comes to the Mountain

Water is amazing here. The Peaks is an oasis, as R.A. Jones put it, following his study of water in the Inner Basin (Jones 1993). The Peaks, like other desert mountains, is also referred to as an altitude oasis (Warner 2004). This is one of the reasons that indigenous people worldwide revere desert mountains. The water that makes its way to the Peaks has been described as some of the purest water in the United States, having very low concentrations of dissolved solids or foreign particles, having come in clouds of evaporated water (Updike 1977). The pristine condition of Inner Basin water was verified recently by an NAU graduate student who found that the water in the Inner Basin Aquifer had very low levels of dissolved solids (Higgins 1998). The water in the Inner Basin Aquifer is also young, perhaps a year or two old (Don Perry, personal communication 2017). By contrast, water from springs around the Peaks ranges in age from about 200 years old in the Lake Mary area to more than 5,000 years old near Wupatki to the east (Bills et al. 2007).

Winter storms typically come from the Pacific Ocean, flowing over the Peaks from the West, summer monsoon storms typically form from moisture that comes from the south, from the Gulf of Mexico, the Gulf of California and the southern Pacific Ocean. Summer storms are initiated by the heating of the southwestern deserts. Moist air, evaporated from the gulf waters, is drawn inland by this heat, and then upward and northward by the increased elevations of the Mogollon Rim, the San Francisco Peaks, and even the Rocky Mountains.

According to local meteorologists Lee Born and Brian Klimowski, the elevated Mogollon Rim is the primary driver in bringing moisture to the Peaks and surrounding area. The Mogollon Rim—as the southern edge of the Colorado Plateau—is a massive, uplifted escarpment that stretches more than 200 miles (322 km) across much of northern Arizona and into the western edge of New Mexico (Fig. 3.10, Plate 3). It is described as an orographic feature, and so like other elevated land masses including mountains, it strongly affects the movement of air. The rim's rise—thousands of feet above the deserts to the south—forces air moisture-laden air masses to rise. As it cools with increased elevation, this cloud-based water vapor condenses and falls as rain or snow over the higher terrain.

During the monsoon season, the Peaks serves as an 'elevated heat source,' so as the Peaks is heated during the day, its warmed surface heats surrounding air, causing it to rise, routinely forming thunderstorms in the summer. All of these features help to draw moisture from air masses that move into this part of Arizona from the southern gulfs and the Pacific Ocean during our monsoon storm season, and even from the central and northern Pacific during the winter months (see below).

So, the influence of the Mogollon Rim and the San Francisco Volcanic Field results in greater moisture over the area (Fig. 5.3, Plate 5). Without the Peaks and the Mogollon Rim, the region's annual precipitation might well be in the single digits, similar to the surrounding deserts. The extent of this blessed moisture zone is about 240 square miles (620 km^2). Other volcanic mountains to the south and west

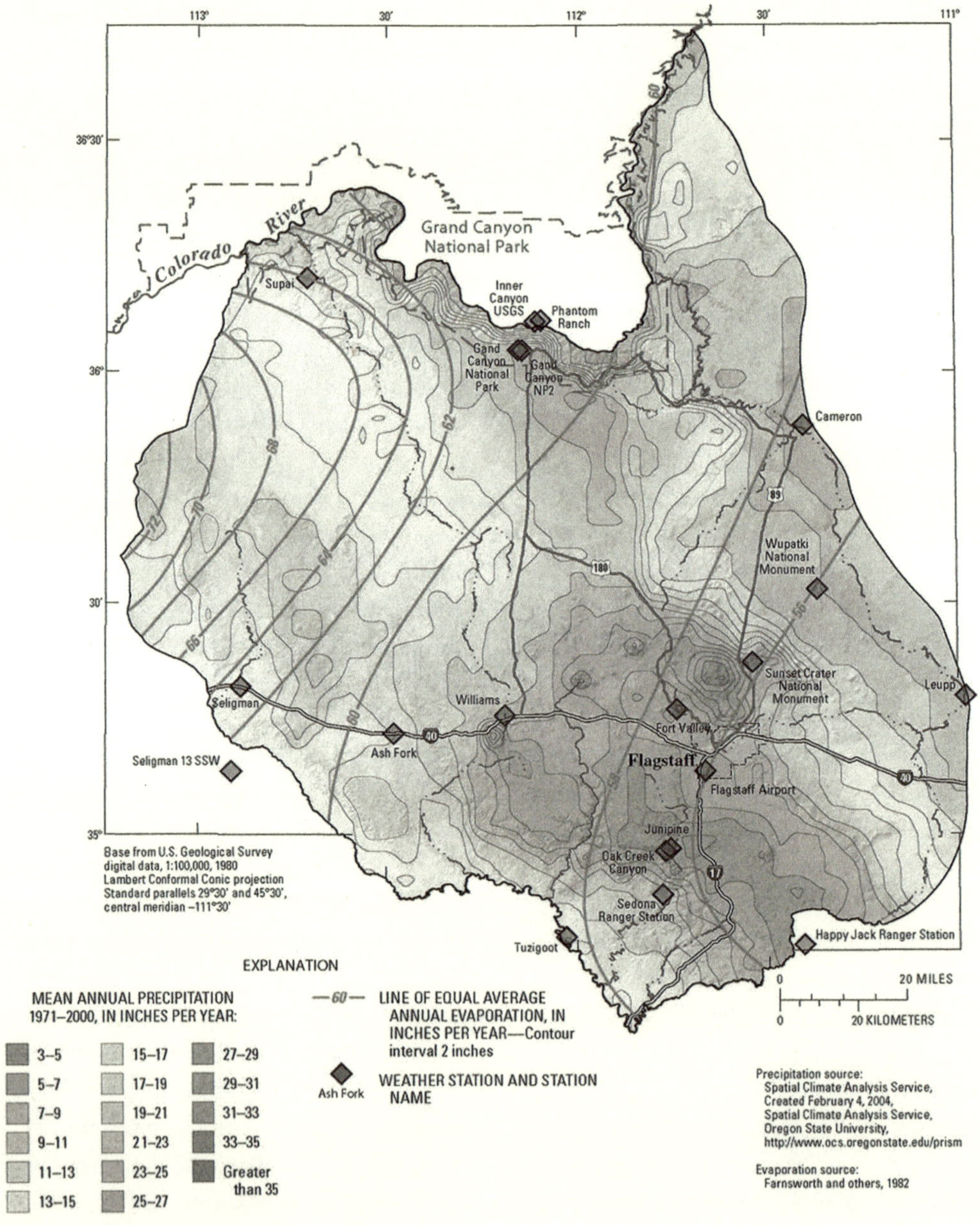

5.3. Average precipitation and evaporation estimates on the Coconino Plateau. Graphic courtesy of Bills et al. (2007) and USGS. Data obtained from Spatial Climate Analysis Service, Oregon State University, 2004. Note how dark sections, indicating higher moisture, are clustered in the SFVF. See Plate 5.

of the Peaks intercept lots of moisture, with Bill Williams, Sitgreaves and Kendrick each receiving more than 30" (760 mm) of precipitation each year on average. The town of Flagstaff, at the base of the Peaks, receives about 21" (530 mm) of annual precipitation on average, benefiting greatly from its position on the Mogollon Rim and its proximity to the Peaks. Meanwhile, the volcanic mountains to the east of the Peaks lie in the rain shadow of the Peaks and the Rim. The town of Leupp AZ, 40 miles east of the Peaks at an elevation of 5,000 feet, only receives 5-7" (180 mm) of precipitation each year; it is in the Cold Desert (Fig. 5.3). Phoenix, a mere 150 miles (241 km) south of the Peaks, receives about 6" (150 mm) of precipitation per year, at an elevation of about 1,000 ft (300 m). So living by a mountain and/or an uplifted plateau seems like a good plan for having access to water.

Precipitation on the Peaks

Just to be clear, precipitation is the total of snowfall and rainfall, of all moisture from the sky. The story of precipitation on the Peaks is still unfolding. Descriptions of climate in the area around the Peaks date back to the mid-1800s, and weather in the Flagstaff area has been measured for more than 100 years (e.g., Cline 1982, Robinson 1913, Sellers & Hill 1985). Formal studies, and particularly those that considered the relationship of precipitation and elevation, started early, too.

Gustav Pearson

Pearson was the first to describe the relationship between elevation and precipitation on the Peaks. He measured precipitation, evaporation, soil moisture and air temperature in 5 different life zones: pinyon-juniper, ponderosa pine, mixed conifer, and spruce-fir zones, and at treeline, along a gradient on the west side of the Peaks, starting in 1917 (Pearson, 1920a, b, 1931). Pearson found a steady increase in precipitation with a climb in elevation, and a decrease in temperature and evaporation with increasing elevation (Table 3.3). He showed that the mixed conifer zone really is the 'sweet spot' on the Peaks, with significant precipitation and more frost-free days—a longer growing season—than even the ponderosa pine zone below (see Mixed conifer section). His experiments should be replicated. He also described how even soils vary across this gradient (see Soils above).

Harold Colton and the Peaks' rain shadows

Director of the Museum of Northern Arizona Harold Colton measured precipitation on the Peaks between 1932 and 1956 (Colton 1958). He was the first to measure precipitation in the Inner Basin. The average amount of precipitation was 33 inches (840 mm) per year in the basin. Summaries for that site were based on only 5 years of data, due to difficulties with snow and access.

Colton's study was the first to detect a rain-shadow effect on the north and east sides of the Peaks; in other words, the north side of the Peaks received less precipitation per year than the south side (15.77 versus 19.16" respectively, 400 vs. 500 mm), and the east side received less than the west side (17.06 vs. 22.57" 430 vs. 600 mm; Colton 1958), as most precipitation is intercepted from incoming storms on the west and south sides of the mountain. This is a study that should be repeated. I noticed that the understory plants are much thinner on the north side of the Peaks, and perhaps the rain shadow effect is influential.

Seasonal patterns of precipitation on the Peaks

Precipitation makes it to the Peaks year-round on average, though most precipitation (73%) arrives biseasonally—in summer monsoon storms (July to September) and in winter snowfall (December to March) (Fig. 5.4). SNOTEL data show that on average, the Peaks receives most precipitation in summer and winter, while the months of May and June are especially dry (see below). The patterns were the same for both sites on the Peaks, though of course, Snowslide Canyon received more precipitation than Fort Valley (Fig. 2.6).

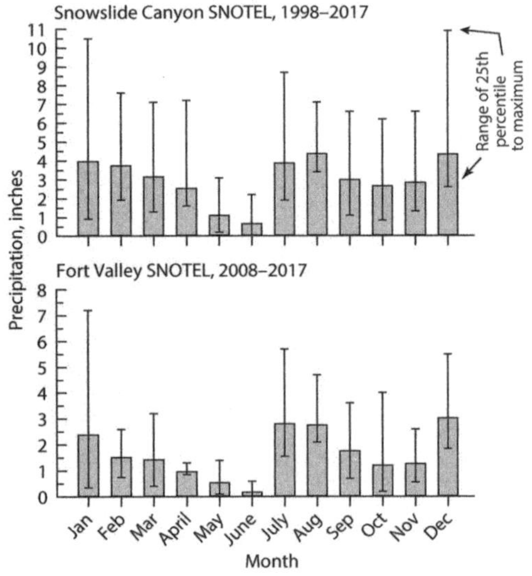

5.4. Annual precipitation patterns at low and high elevations on the Peaks, based on SNOTEL data for the two sites. Courtesy of Richard Hereford, USGS.

Long-term precipitation patterns on the Peaks

Annual precipitation atop the Peaks and the Inner Basin averages 41.3" (1,050 mm, Fig. 5.2), which is double that of Fort Valley and Flagstaff, which are lower in elevation. Fort Valley is 5,292 feet below Humphreys Peak. The driest year on the Peaks was 2002 (22.2" or 563 mm), followed by 1956 (19.7" or 500 mm). The wettest year on the Peaks was in 1965, with 70.8" (1800 mm, followed by 1905 with 64.7" (1,640 mm). Precipitation varies widely from year to year.

Why precipitation varies

This large annual variability relates to the high elevation of the Peaks in the area, the orographic effect of the mountain (see above), and variable changing states of the El Niño-Southern Oscillation (ENSO). A see-saw pattern of atmosphere pressure between the eastern and the western Pacific Ocean (the Southern Oscillation) and corresponding changes in sea surface temperature identify El Niño and La Niña seasons. Increased sea-surface temperature of the eastern Pacific Ocean (El Niño) and vice versa (La Niña) are signs of ENSO activity that affect the Colorado Plateau (Hereford 2002) and Southwest climate. Although the effect is variable, during El Niño the weather here can be unusually wet. On the Peaks, El Niño noticeably enhanced precipitation in 1905-06, 1940-41, 1964-65, 1978-79, 1983-84, and 1992-93. The flip side of El Niño is La Niña, which more reliably favors decreased, below-average precipitation in the Southwest and on the Peaks; notably dry years are 1909-10, 1950-51, 1955-56, 1976-77, 1988-89, and 2001-02 (Fig. 5.2).

The long-term patterns of alternating wet and dry conditions roughly correspond to those throughout most of the Southwest (Hereford et al. 2014). The early part of the precipitation history is relatively dry until about 1905 when an abrupt and persistent pattern of variable, typically above average precipitation began, and lasted until the late 1940s (Fig. 5.2). Dendroclimatologists recognize this period as the Early 20th

Century Pluvial (Woodhouse et al. 2005, Salzer & Kipfmueller 2005), which was the wettest episode of the preceding 1,200 years. The wet period or pluvial was followed in the early 1940s by the Mid-20th Century Drought. This long-term multidecadal precipitation variability on the Peaks is likely related to the Pacific Decadal Oscillation (PDO), which affects precipitation over 15-30 year periods. The PDO is an index of oceanic temperatures in the Eastern Pacific Ocean. In its warm phase, climate in the Southwest tends to be relatively wet, while in its cool phase, precipitation is usually at or below normal (Hereford et al. 2006).

SNOTEL: year round precipitation patterns

Today, there are two permanent automated weather stations that measure precipitation and temperature year round on the Peaks! These state of the art weather stations provide the most accurate measures of precipitation and temperature today, and will provide the most important measures going forward. A SNOTEL system was established at Snowslide Springs in the Inner Basin in 1996, and one was established at Fort Valley in 2004 (Fig. 2.6). It sure would be nice to have more of them on the Peaks; they could describe where the most precipitation falls on the mountain, describe rain shadows, follow temperature change, etc.

SNOTEL data show large differences in the amount of precipitation between the low elevation and higher elevation sites and between years (Fig. 5.5). Precipitation at Snowslide Canyon ranged between about 20 inches and nearly 50 inches per year (500 to 1300 mm). These data also show that 2002 was one of the driest years in much of the Southwest during the 21st century.

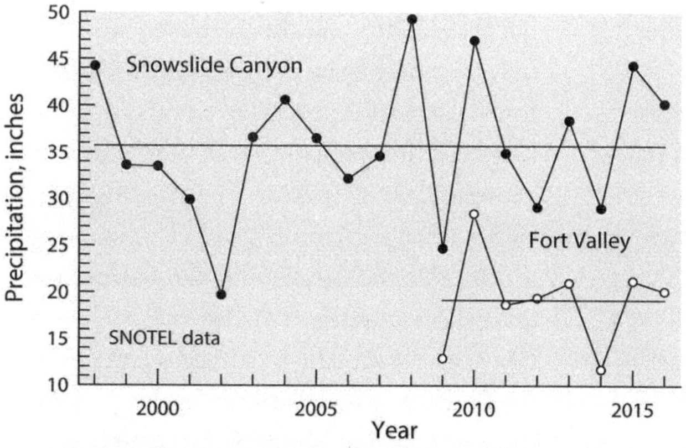

5.5. Annual precipitation at Fort Valley and Snowslide Canyon, using SNOTEL data, 1998 to 2016. Note the substantial difference in precipitation amounts between the low elevation Fort Valley site, and the high elevation Snowslide Canyon site. Courtesy of Richard Hereford, USGS.

Snowfall and elevation on the Peaks

The snow course technique has measured snowfall at 4 sites along an elevational gradient on the Peaks (Fig. 5.6, Plate 6). Data collection began at Fort Valley in 1947, and at Snowbowl in 1965, and in the Inner Basin starting in 1968. As mentioned, it is snowfall that feeds the soil, deep rooted plants and regional aquifers, so it is a very important measure (Higgins 1998, Bills et al. 2007). A snow course is a permanent transect along which snow depth and water content are determined by extracting cores of snow (Jones 1993). These data were collected one to two times per month through the winter at each site. The snow course technique measures only winter precipitation, and was expected to underestimate snowfall by making measurements only once or twice a month (Jones 1993). However, snow course data and SNOTEL data are well-correlated, as it turns out, which

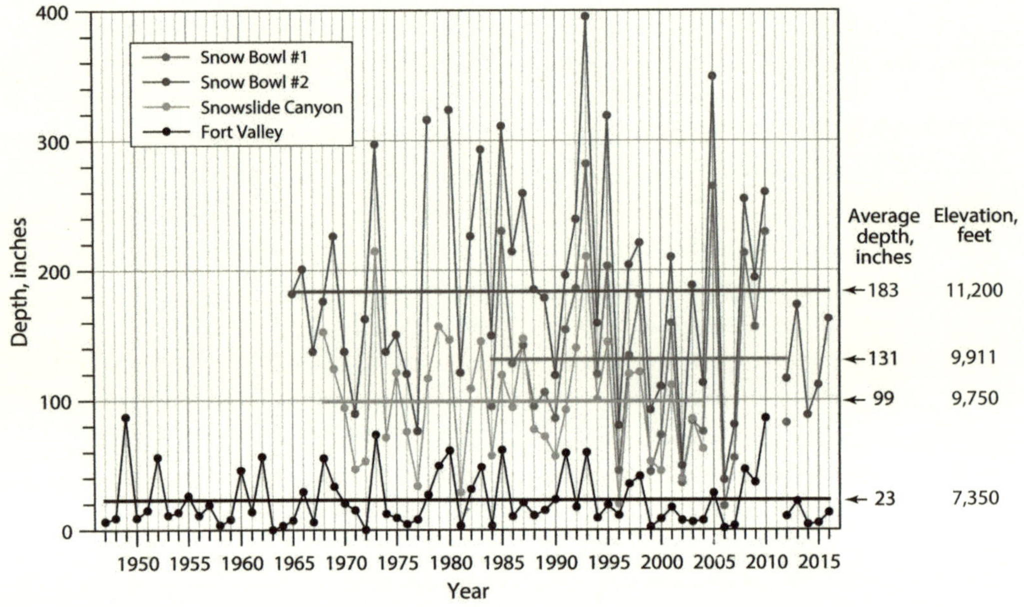

5.6. Annual snowpack depths, with averages, across elevations on the Peaks, based on snow course data for Fort Valley, Snowbowl 1 & 2, and Snowslide Canyon in the Inner Basin. Courtesy of Richard Hereford, USGS. See Plate 6.

helps to validate the longer-term snow course data (Richard Hereford, personal communication, 2017).

The Inner Basin Aquifer

Another measure of water on the mountain is that which lands in the Inner Basin. And a lot of that drains into the basin's glacial moraines. The City of Flagstaff developed a water collection system in the Inner Basin, and measures the amount that it collects from the moraines and springs through the year. And it can be quite substantial!

The glacial moraines in the Inner Basin collectively represent a perched aquifer called the Inner Basin Aquifer—the largest perched aquifer on the mountain, meaning that it can hold large volumes of water (Fig. 3.1; see glaciers above). These moraines with their glacio-fluvial sediment are estimated to have a massive volume of nearly 9 billion ft^3 (255,000,000 m^3; Higgins 1998). For some perspective, the Core Ridge terminal moraine alone is over 500 feet high in the mid-basin; so the debris pushed down the Inner Basin ahead of glaciers got to be this high above the basin's floor, and the glaciers themselves may have been even taller (Higgins 1998). Inner Basin springs issue from this aquifer (Fig. 3.1). It is recharged each year largely by snowmelt, with rainwater contributing little (Jones 1993, Higgins 1998).

In 1899, the City of Flagstaff began to divert spring water in the Inner Basin for the city of Flagstaff, which lies just southwest of the Peaks. The installation of a clay pipeline from Jack Spring, about half way up the basin, came first, with the pipeline extending down the waterline road, on the eastern flanks of Schultz Peak, to large holding tanks on the north edge of Flagstaff. Once operational, a gauge determined that this source produced 150,000 gallons of water (568 m^3) every 24 hours (Higgins 1998). Between 1899 and 1960, water from six more springs, including Snowslide, Flagstaff, Doyle, Raspberry, Bear Paw and Beard springs, was diverted for municipal use (Fig. 3.1). At this time, nearly 95% of Flagstaff's water came from the Inner Basin and 5% from the north side of the mountain, from a spring that was diverted in Abineau Canyon. Several springs in the Schultz Pass area also fed the pipeline to Flagstaff (Higgins 1998).

After securing water sources from Lake Mary and the Woody Mountain well fields southwest of the Peaks, the City of Flagstaff, in search of more water, began drilling wells directly into the Inner Basin Aquifer—the glacial moraines—in the 1960s. Ultimately, 15 wells were drilled, and today 3 of them provide water to Flagstaff. Eight of the wells extended to depths between 170 to 500 ft (52 to 152 m), without reaching the aquifer's bottom. Most of the wells were exploratory, some were drilled too deeply and broke through the Inner Basin floor, causing water to disappear down into the mountain, perhaps draining into the Coconino Aquifer. One of the wells was found to be artesian during initial pumping tests (Higgins 1998). It's hard to think about the impact this major disturbance had on the Inner Basin.

As Ronald Jones pointed out, the city water production data provide only a limited measure of actual precipitation, because the pipeline carrying the water from the Inner Basin has a finite carrying capacity, the well pumps don't always run, the pipeline leaks *and* the system is shut down for repairs periodically. That said, because this is a long-term measure of year round water in the basin, I felt it was worth presenting, to provide *some* measure of the amount of water the Inner Basin receives, conservative measure that it may be.

Between 1970 and 1991, Inner Basin water production for the city averaged about 332 million gallons per year (1,256 m^3). During this period, annual output ranged from a high of 580 million gallons (2,200,000 m^3) in 1980 to a low of 56 million gallons (212,000 m^3) in 1991 (Jones 1993). Although 1991 was an above average snow year (Fig. 5.2, 5.6), the following spring was dry and windy, which led to tremendous snow loss through sublimation (Jones 1993; see sublimation below).

Another study of Inner Basin precipitation—also based on the city water collection system—estimated that in 1997 the city water system captured 178 million gallons (674,000 m^3) of water, indicating a low precipitation year. It also estimated that 755 million gallons (2,860,000 m^3) recharged the aquifer that year (Higgins 1998).

The USGS considered the amount of water in all springs and streams in the Inner Basin, including those not collected into the city system (Bills et al. 2007). The average estimated base-flow from the Inner Basin springs alone is 471 million gallons (1,800,000 m^3) of water per year! The annual base-flow estimate from springs for the entire San Francisco Volcanic Field is 1.5 billion gallons (5,700,000 m^3), meaning that nearly 30% the estimated spring flow in the entire San Francisco Volcanic Field comes from the Inner Basin alone (Bills et al. 2007)!

In an effort to calculate the water budget for the *entire* mountain, Higgins estimated that the recharge from the entire mountain could be 5 times greater than that estimated for the Inner Basin alone (Higgins 1998). By this measure, based on the 1997 estimated

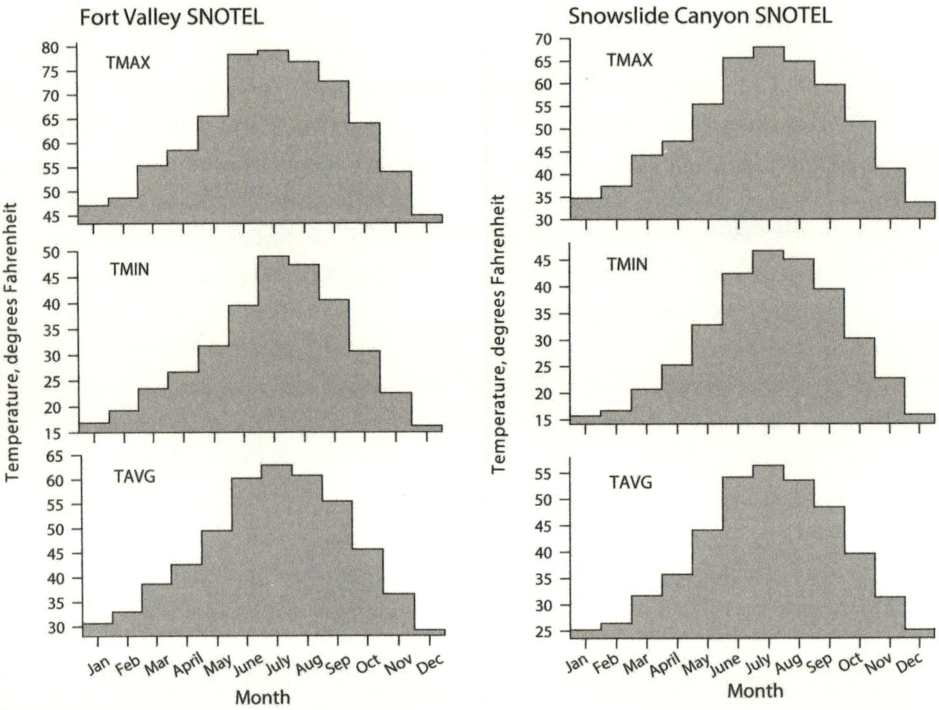

5.7. Annual temperature cycle at Fort Valley and Snowslide Canyon, using SNOTEL data showing monthly averages of maximum (TMAX), minimum (TMIN), and average temperatures (TAVG). Note that temperatures peak in July. Courtesy of Richard Hereford, USGS.

recharge of 763 million gallons (3,000,000 m^3) of the Inner Basin Aquifer, nearly 4 billion gallons (15,000,000 m^3) may have fallen on the entire mountain at least briefly! Talk about a mountain oasis.

Seasonal temperatures

Average temperatures through the seasons vary greatly on the Peaks, by nearly 40 °F (4 °C) over the course of a year. And, of course, temperatures are much higher at Fort Valley than at Snowslide Canyon (Fig. 5.7).

Long-term temperature patterns on the Peaks

This long-term record shows a significant pattern of warming near Humphreys Peak and the Inner Basin between 1895 and 2015 (Fig. 5.1; Prism Climate Group, http://prism.oregonstate.edu). Maximum and average temperatures increased significantly by 2.4 and 0.9 degrees per century, respectively. This pattern matches global warming trends (Richard Hereford, personal communication 2017). Average minimum temperatures have not increased significantly, though minimum temperatures are increasing. More short-term data sets from SNOTEL weather stations on the Peaks show a similar warming trend at both Fort Valley and Snowslide Canyon (Fig. 5.7).

Average annual temperatures and minimum temperatures have increased significantly at both lower elevation Fort Valley, and higher elevation Snowslide Springs (Fig. 5.8). Although maximum average temperatures were always higher at Fort Valley, minimum temperatures were similar at both elevations. This latter

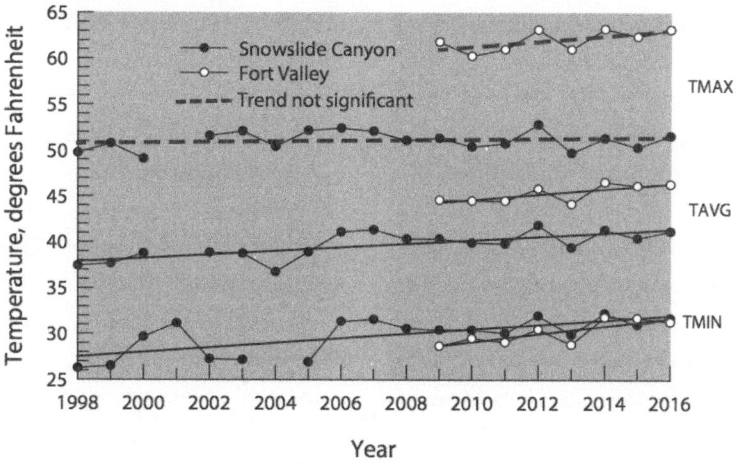

5.8. Average annual maximum (TMAX), minimum (TMIN), and average temperatures (TAVG) at Snowslide Canyon and Fort Valley using SNOTEL data, 1998 to 2016. Note the statistically significant increase in minimum and average temperatures. Courtesy of Richard Hereford, USGS.

surely relates to the fact that cold air descends down the Peaks at night, making the ponderosa pine zone in Fort Valley colder than the mixed conifer zone above it (Pearson 1931, see ponderosa pine section).

Where The Water Goes

So the Peaks collects an amazing amount of water. However, most of it goes away—through leakage or infiltration down into the mountain and through evaporation. This is where things get kind of weird, and shows how incredibly dynamic the life of water in this southwestern mountain is.

Leakage or infiltration

Patrick Higgins' study of Inner Basin precipitation and discharge suggests that up to 80% of the of the water arriving in the Inner Basin Aquifer is not collected in the city lines (Higgins 1998)! So the amount of water that falls on the Inner Basin is actually far, far larger than either recharge or diversion measures suggest! The very fractured nature of this composite volcano allows for enormous leakage of water from the aquifer down into the mountain.

As water saturates the base of the Inner Basin Aquifer, it is driven deeper into the mountain, although it is slowed a bit by an 'old colluvium' layer underlying the moraines (Higgins 1998). It is suggested that without this layer, little water would accumulate in the aquifer. Some of the moisture that lands on the Peaks seeps into its many fractures, with lots of conduits to carry water. Rainfall and snowmelt drain into perched, shallow aquifers all over the mountain, often issuing out through springs, and percolate down into the underlying Coconino Aquifer, and possibly flow through interior conduits—all the way to Kendrick Mountain and to the Little Colorado to the northeast (see Section 2 on Sugarloaf Mtn.; Higgins 1998).

Evaporation and sublimation

Water loss to evaporation in summer and evaposublimation in winter is enormous! According to Bills et al. (2007), the average loss of moisture via evaporation and sublimation in the vicinity of the Peaks each year is nearly 58" (1470 mm). Given that the area gets 20-40" (500 -1000 mm) of rain and snow annually, well, that's really hard to imagine (Fig. 5.3, Plate 5; Bills et al. 2007).

In summer, rainfall runs off quickly, and because it is warm out, it evaporates at a rate that ultimately exceeds annual precipitation levels. Water evaporates from water surfaces such as lakes and streams, and from as much as a foot deep in soil, and it is transpired by plants. Plants evapotranspire or breathe out water into the atmosphere through pores or stomates in their leaves. They do this when they are pulling in CO_2 from the air, to make carbohydrates, or when they are heated, whereby it becomes a means of heat and water loss. They do a lot of both in the growing season, and represent an important source of evaporation.

In winter, sublimation results as moisture vaporizes directly from snow—from frozen water, when temperatures are at or below freezing. Winter on the Peaks is marked by frequent periods of dry weather with lots of sunshine. Windy conditions often occur due to the high elevation of the Peaks relative to the surrounding landscape. These factors and the low latitude of the Peaks all cause a significant portion of its snow to sublimate. Sublimation or loss of snow to vapor is greater in dry climates than in wetter climates (Etter 2006).

There are few studies that specifically address sublimation on the Peaks, though both of the precipitation studies above suggested that sublimation was very important in water loss in the Inner Basin, and cited the need for more research (Jones 1993, Higgins 1998).

NAU Hydrologist Chuck Avery et al. (1993) suggested that the amount of snow lost to sublimation in the Inner Basin could reach 50% or more, possibly even equaling an average winter's snow accumulation. The one sublimation study on the Peaks found 43% of the snow level was lost through sublimation on the western flanks of the Peaks (Froyland 2012)!

A study of sublimation at Wing Mountain, near the Peaks, showed how complex that process can be (Etter 2006). Etter considered sublimation on an exposed treeless slope, on a treeless slope in shade from a nearby forest, and a slope with a dense forest canopy. In early winter, the forested slope experienced the greatest sublimation. The branches of the conifers absorbed heat during the day, and released it at night, resulting in sublimation into the night. Later in winter, sublimation on the open slope exceeded that in the forested area, due to greater heat absorption as the sun was higher in the sky.

Jones (1993) provided an example of how sublimation in the Inner Basin works: he described a very wet, late winter storm in the Inner Basin in 1991, and yet because April and May were very dry and windy, considerable snow was lost to sublimation, to the extent that snow water did not recharge the aquifer that year. And that was in a year that received a decent amount of snow! Jones advocated for greater use of SNOTEL weather stations on the Peaks, to measure both evaporation and sublimation.

So, if water measurements are made *after* so much of the precipitation that lands on the Peaks evaporates, then the amount that at least initially reaches the mountain is even greater, more than double the above estimates.

Water from springs

The Springs Stewardship Institute, based at the Museum of Northern Arizona, has mapped 22 springs on the Peaks, and an additional 14 springs nearby (Ledbetter et al. 2018). Thanks to this program, springs in many parts of the United States are starting to receive much needed recognition for their ecological importance. This goes for the springs on the Peaks, as well. Although most springs occur in the Inner Basin, and are associated with the aquifer (Fig. 3.1), a few are found on nearly all sides of the outer flanks of the Peaks. A survey of springs on the Peaks determined that 83% were severely impaired or no longer functioning as springs. They have been somewhat to entirely altered; most commonly, alterations were geared to provide water for human use and for livestock: 7 had berms or dams created to provide water tanks, while the water at 9 springs was piped away, leaving 6 of them entirely dry. Most springs have been heavily disturbed at some point for the purpose of diverting their water.

Snowslide Spring almost functions as an undisturbed spring, as its outflow is diverted quite a bit downstream from its source. This is an exceptional spring, with its large population of the beautiful Parry's primrose, geum, buttercup, sedge, etc. (Fig. 5.9, Plate 40).

Big Leroux Spring is fairly-well studied, and a group in nearby Flagstaff, Friends of the Rio de Flag, is advocating for its protection and the restoration of its channel and ecological integrity. Leroux Spring is a perennial spring that has provided water to travelers for hundreds or even thousands of years. It occurs on the southwest side of the

5.9. Parry's primrose spilling down Snowslide Canyon below Snowslide Spring.

mountain near its base. The Sitgreaves party, led by regional explorer, guide and trapper Antione Leroux, stayed there in 1851, along with surgeon-naturalist S.W. Woodhouse (see Reptiles and Amphibians above). The spring was beautifully described in 1857 by Lt. Edward F. Beale as "...one of transparent, sparkling water [that] bursts out of the side of the mountain and runs gurgling down for a quarter of a mile, where it loses itself in the valley." Beale had great regard for Big Leroux Spring, and oriented the Beale wagon road near the spring (Monroe et al. 2013). This spring is largely or nearly perennial, though its flow has decreased in recent years. E.C. Martin measured flow at Leroux Spring in the 1960s, and ultimately described it as producing about 40,000 gallons per day (151 m^3), with little fluctuation regardless of precipitation levels or season (Martin 1969). According to Susan Olberding's book on the history of Fort Valley, Big Leroux Spring released 14,400 gallons of water per day (54 m^3, Fig. 2.6) in a 1981 engineering report (Olberding 2002).

In 2003, geologist Stephen Monroe started measuring flow at Leroux Spring, at the point where it was diverted into a pipeline. He found the flow to be strong, and with Friends of the Rio de Flag, proposed to the U.S. Forest Service that some of the spring's diverted flow be returned to its natural course, a drainage which once housed a diverse riparian community of plants and animals (Monroe et al. 2013). They described the spring as having had one of the largest flows among the Peaks' springs. In 2013, the valves in the spring box were adjusted, releasing 40% of the spring's output back into its channel, thereby beginning a restoration of the riparian zone below the spring!

Nearby Little Leroux Spring's flow is more affected by season and precipitation levels, producing 900 gallons per day (3.5 m^3) in spring in a dry year, and 8,300 per day (31 m^3) in spring in a wet year (Martin 1969).

Little Spring, to the north, is another lush area on the Peaks, and its easy to see why the Merriam party established their base camp there in 1889 (Fig. 3.18). It was much modified at one point to provide water for livestock. It drains from a pipe that runs often or perennially, given the green nature of the place. This place has a great diversity of plants and birds.

Runoff

I have made a drainage map for the Peaks based on Bills et al.'s (2007) general drainage descriptions for the mountain, and on drainage patterns revealed during the epic flooding on the east side of the Peaks following the 2010 Schultz Fire. This is only a first attempt at such a map, since there is nothing else available; it will surely be revised. At times, and especially during the Pleistocene, most water draining from Peaks made its way east to the Little Colorado River, which, of course, pours into the Colorado River in Grand Canyon. There is one exception though: water coming down Hart Prairie on the western side of the mountain actually heads southwest all the way to Sycamore Canyon and from there to the Verde River (Fig. 5.10, Bills et al. 2007). Runoff probably hasn't made it to the Verde River in some time. Bills et al. (2007) describe the runoff from the southeastern and eastern flanks of the Peaks as draining into Walnut Creek and San Francisco Wash, as it makes its way to the Little Colorado. Flooding following the Schultz Fire in 2010 showed that water draining down

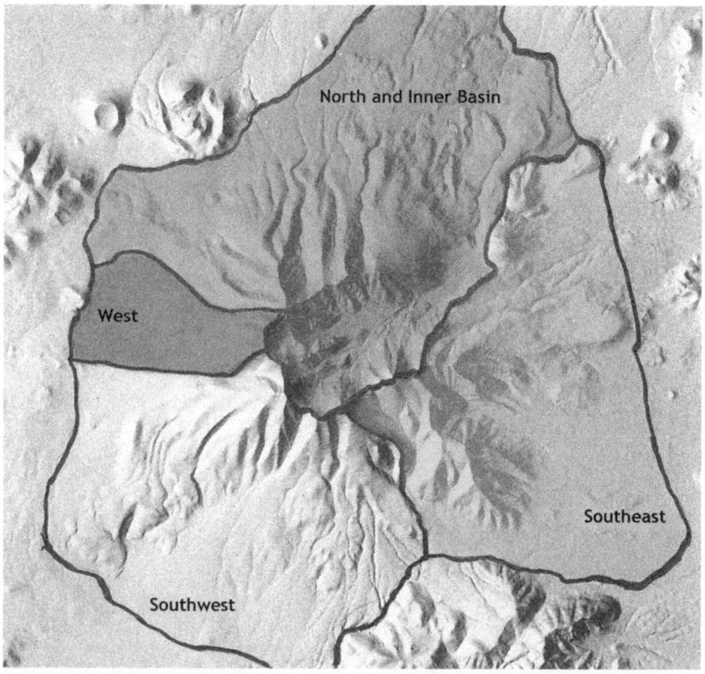

5.10. Approximate drainage basins for the Peaks. They include: Northern and Inner Basin drainage into Deadman Wash and on to the Little Colorado River, Western flow to Sycamore Canyon and the Verde Valley, Southwest into the Rio de Flag, and Southeast to the Little Colorado River.

Weatherford Canyon and the west flanks of Schultz Peak made its way to the east, along with water coming down the eastern flanks of Schultz Peak. This indicates a drainage divide between Weatherford Canyon and upper Schultz Creek. The headwaters for Schultz Creek appear to be at Orion Spring or just above it.

Peaks watersheds

Southwest side of the mountain: this area drains the south and southwest side of the Peaks, from Schultz Creek through Freidlein Prairie, all draining into the Rio de Flag; **West side of the mountain**: a small watershed drains the west side of the Peaks through the Hart Prairie area into Volunteer Wash, which forms at the base of Fern Mountain, and heads southwest to Sycamore Canyon and the Verde River (Fig.2.6; Bills et al. 2007). This drainage crosses volcanic rocks and then cuts deeply into the Paleozoic rocks of the Mogollon Rim. **North side and Inner Basin**: Deadman Wash drains the north side of the Peaks *and* the Inner Basin. This wash originates near the northwest edge of the Peaks, approximately 6 miles (9.6 km) east of Kendrick Mountain. **Southeast side of the mountain**: Schultz Peak, Weatherford Canyon, the northeast side of Dry Lake Hills and Elden Mountain constitute this drainage basin. Flooding following the Schultz Fire in 2010 made it very clear that Schultz Peak drains widely into the flats on the east side of the Peaks, especially Black Bill Park and Doney Park.

There may have been yet another drainage route out of the Inner Basin before the formation of Sugarloaf Mountain 91,000 years ago. The eastern edge of the Inner Basin, where Sugarloaf is now, probably provided a route for glacial outwash all the way to the Little Colorado, and indeed, there is glacial outwash in Bonito Park and near the base of Sunset Crater (Updike 1977).

These days the Inner Basin occasionally drains down an unnamed tributary of Deadman Wash. This canyon is several hundred feet deep. In 1983, NAU geologist Richard Holm, while mapping the lava flows on the north side of the Peaks, recalled seeing water flowing from the Inner Basin and running across Forest Service road 418 along the northeastern base of the mountain.

In his 1904 report on the Peaks, John Leiberg described 'Smith Creek' as a perennial though 'insignificant' runlet originating in the western end of the Inner Basin. He said it flowed for about a mile and a half before sinking, presumably into the moraines in the mid-basin. He referred to this creek as a major source of water for the City of Flagstaff (Leiberg et al. 1904). It might have originated up Snowslide Canyon, which runs through much of the early summer, though most of it is diverted now.

Water draining down the Peaks' outer slopes flows into many shallow basins or parks that occur around the mountain, including Bonito Park, Lenox Park, Black Bill Park, and Doney Park (Feth 1953). Doney and Black Bill Parks flow into the Rio de Flag, though the drainage pattern is poorly understood (Richard Hereford, personal communication, 2018). A study by Cosner (1962), found alluvium to a depth of more than 300 ft (100 m) in Doney Park.

Fort Valley is a valley of the Rio de Flag (Fig. 2.6). The Rio flows through Fort Valley, so not surprisingly, Fort Valley has a very high water table, with water historically found within 3 feet of the surface (Olberding 2002). Populations of water-loving plants like geranium-leaf delphinium and the rare western porterella (*Porterella carnosula*) can be found in the wet soils there.

Runoff, Floods and Erosion

I've said it before and I'll say it again, I just love the thought of glacial outwash flowing out of the Inner Basin and the northern drainages of the Peaks, into Deadman Wash, and making it all the way to the Little Colorado River during and after the Pleistocene (Fig. 2.1)! Glacial meltwater moved volcanic rocks all the way to the Little Colorado River along this route during the Pleistocene. The large rocks in Deadman Wash gravels in Wupatki Basin are identical to the volcanic rocks in the Inner Basin (Ulrich et al. 1984, Richard Hereford, personal communication, 2017).

Since the end of the Pleistocene, the Rio de Flag and one of its tributaries, Schultz Creek, are the only drainages running out of the Peaks to any degree. Neither is now perennial. Schultz Creek drains a large area, from the western edge of Weatherford Canyon west towards Friedlein Prairie and the backside of most of Dry Lake Hills. The headwaters of the Rio were fed by Leroux and Veit Springs at one time.

There have been many big floods down the Rio de Flag. The records are mostly for the town of Flagstaff, to the south. Between 1888 and 1995, there were at least 19 major floods, with flows reaching between 500 and 1,200 cubic feet per second. That equals one big flood about every 5 years out of the west side of the Peaks. A partial list of recorded large floods includes those in 1888, 1896, 1903, 1905, 1916, 1920, 1923, 1937, 1938, 1950, 1963, 1966, 1973, 1979, 1983, 1988, 1990, 1993, and 1995. Many of these were high snowpack years on the Peaks (Fig. 5.2, 5.6). Roads, bridges, and buildings in downtown Flagstaff and south Flagstaff were inundated with floodwater during many of these events.

The largest flood-producing area along the Rio de Flag is on the south side of the Peaks between Schultz Creek and Fort Valley, while Schultz Creek drains a much smaller area (Fig. 2.6). USGS geomorphologist Richard Hereford suggests that the Rio could deliver a *really* devastating flood, if the forests in its headwaters were destroyed by fire. Catastrophic flooding from the headwaters of the Rio de Flag would devastate the west side of Flagstaff, which was developed on the Rio's presettlement floodplain. That's what happened on the east side of the Peaks following the 2010 Schultz Fire (see below).

The Rio de Flag has a really convoluted path to the Little Colorado River to the east. It heads east from Agassiz Peak and joins San Francisco Wash after about 24 miles (37 km), then San Francisco Wash feeds into Wildcat Canyon (north of Cosnino), then that drains into Padre Canyon, which feeds into Canyon Diablo, which feeds into the Little Colorado River, all that over the course of about 52 miles (84 km; Richard Hereford, personal communication 2017).

As mentioned in the Holocene description above, after nearly 6,000 years of accumulating soil in its channel, Schultz Creek started running again, becoming erosional about 100 years ago (Stempniewicz 2014). Erosion or incision in Schultz Creek, on the south edge of the Peaks, began in the late 1800s, with a 100-year flood occurring in Flagstaff in 1923 (Stempniewicz 2014). This date corresponds to an increase in effective precipitation from 1907-1926, and an increase in the frequency of large floods, and subsequent erosion across northern Arizona (Hereford 1984, 2002, Salzer & Kipfmueller 2005, Stempniewicz 2014). Climate played a role in the flooding and erosion at the turn of the 19th century, along with human impacts (Masek-Lopez and Springer, 2002). Overgrazing and drought in the mid-1890s significantly reduced vegetative cover, resulting in increased runoff and erosion and severe downcutting of streams (Hereford 1984).

There are alluvial terraces in every major canyon on the Peaks, whether they were glaciated or not; they are best formed in the canyons on the north side of the mountain (Philomena, White Horse, Abineau, Reese and Bear Jaw). Some are up to 50 ft (15 m) above present drainage levels. These were formed when enough water was flowing to carry and deposit rocks and sediment in terraces along the stream channels within the canyons. Many of these deposits are related to the Pleistocene, when a lot of water flowed at times! Today, most of the erosion and deposition on the Peaks occurs during short periods under high discharge rates (Updike 1977). In this regard the fluvial or erosional processes on the Peaks are characteristic of many arid or semi-arid environments in the Southwest. There are three types of water-borne deposit: valley-fill sediments in essentially all canyons, alluvial fans below the mouth of some canyons, and basin fill deposits in the lowlands adjacent to the Peaks (Updike 1977).

In a 1936 Museum of Northern Arizona Notes, L.F. Brady described an erosional event where Schultz Creek meets the Rio de Flag (Brady 1936). He described the junction as an open valley, a grassy flat with numerous shrubs and a scattering of very large pines. It was traversed by a small stream, though there was no arroyo or deep drainage. During 1886 and 1887 there was a logging road through the valley. By 1900 a definite arroyo had formed, though it was only 20 ft (6 m) wide. By 1908, a bridge was needed to cross the drainage, due to continued erosion. Measures of erosion at a ranch along the drainage estimated a loss of nearly half a million cubic feet of soil in less than a quarter mile since 1910.

Up near Leroux Spring, in the mid 1920s, Harold Colton described a major summer cloudburst that carried boulder-sized rocks down the flanks of the mountain into resident Buck Taylor's cornfield, rendering it unusable thereafter (Colton 1958).

The Schultz Fire and subsequent flooding

Major flooding occurred after the eastern half of the Peaks burned in the Schultz Fire in June 2010, scorching more than 15,000 acres (6,070 ha; Fig. 3.25, see Fire above). Many of the burned slopes were extremely steep. Extraordinary monsoon storms followed the intense fire that summer, and resulted in widespread flooding that destroyed a lot of the city's water pipeline on the Waterline Road, forced the evacuation of hundreds of homes and sent torrential sediment and ash-laden floods out up to 4 miles (6.4 km) from the burn site. The flood made it all the way to Doney Park, like it did at the end of the Pleistocene, although then it kept on all the way to the Little Colorado River.

The runoff from the burned slopes flowed from nearly 10,000 feet to about 7,000 ft (3,000 m to 2,000 m), scouring thin-soil slopes, deep-soil deposits and then flood plains or debris fans in the flatter area at the bottom. This intense fire caused the soil to become 'hydrophobic' or kind of sealed, so that it could no longer absorb water, which led to even more runoff. Distributing straw and seeds on the upper slopes began soon after the rains subsided, to stabilize the slopes and encourage plant growth to slow erosion in the steepest places. This was done several times and eventually did help to stem erosion.

Various agencies and regional scientists sought to reduce erosion on steep slopes with an enormous mulching effort, and then changed the entrenched channels back into floodplains, broadening them so that floodwater would again spread more widely across them, depositing sediment and water more broadly, which also allowed for more water to be absorbed. They used this treatment in the three most challenging drainages above the flooded neighborhoods, and it has worked. Natural Channel Design, in Flagstaff, AZ, and other regional environmental design companies helped the floodplains to heal themselves. Flooding from the burn area has stopped, but erosion will continue for some time, especially on the higher, steeper slopes. Regional scientists, Coconino County, and the Coconino National Forest came together with some really creative techniques that are healing the damaged landscape, and which will minimize future flooding or damage to downslope neighborhoods.

Its interesting to watch the clonal species like aspen and bitter chokecherry holding down the soil on the burn site these days, along with the help of fields of grasses and flowers (Fig. 3.25).

— SECTION 6 —

More Human Impacts

The Saga Of Snowbowl

This is a bit of history of skiing and snowmaking and ambition on the Peaks.

People were skiing on Hart Prairie on the west flank of the Peaks as early as the 1930s and people enjoy this ski area today. By 1941, a ski lodge up above Hart Prairie was built, a permit was granted to operate a ski area by the U.S. Forest Service, and a dirt road was constructed by the Civilian Conservation Corps.

As the ski area was taking off, its owner, Summit Properties Corporation, purchased 327 acres (132 ha) of land on Hart Prairie, and proposed to develop a ski village, replete with shops, restaurants, lodges, golf courses, and a gondola to carry people up to the ski area. They also sought a rezoning of the land, to accommodate 971 residential lots, in addition to the commercial village, and began to express interest in snowmaking at the ski area. It was estimated that such a development would require 15,000,000 gallons of water annually. In 1969, Summit applied to the Coconino County Planning and Zoning Commission to rezone the land. This proposal marked the beginning of decades of development proposals, including rezoning applications, appeals, and lawsuits opposed to the development which went all the way to the U.S. Supreme Court—all related to the development of a skiing industry on the Peaks. Initial lawsuits were filed by Hart Prairie residents Jean and Richard Wilson and by the Hopi and Navajo Tribes, on behalf of preserving Hart Prairie initially, and, ultimately, the San Francisco Peaks. Continued efforts to develop a commercial enterprise on the western slopes of the Peaks had a strong negative effect on many local and regional residents, and galvanized considerable public opinion against such efforts. The 'Save the Peaks' movement came together, and numerous benefits were held to increase awareness.

The Hart Prairie Planning Area consisted of 3,454 acres (1,400 ha) of private land

and 46,466 acres (18,800 ha) of public land, managed primarily by Coconino National Forest. In 1975, Coconino County estimated that if all private acreage was developed, it would result in 9,672 residential lots, with a projected population of more than 20,000 people! The County concluded that such a development would not be sustainable.

In addition to the growing public opposition to the Hart Prairie development, other important efforts were set in motion which ultimately helped prevent an enormous development in the western foothills of the Peaks. During the late 1970s and the 1980s, Coconino County and Coconino National Forest negotiated land exchanges through which privately held land on the Peaks became part of Coconino National Forest, thus consolidating more of the forest, in exchange for National Forest parcels elsewhere in the state. Furthermore, in 1981, Coconino County adopted a new zoning ordinance that changed the zoning of County land on the Peaks from 1 residence per acre to 1 residence per 10 acres (4 ha). This resulted in a tenfold reduction in the number of permissible residences in the Hart Prairie area. Efforts by Coconino County and the Forest Service helped to limit commercial development on Hart Prairie. In the end, the proposed large-scale commercial development of Hart Prairie, on the western flanks of the Peaks, was thwarted.

Snowbowl Ski Area continued to request expansion of the ski area from Coconino National Forest, but it was its request to use Flagstaff's reclaimed water or effluent for snowmaking that redefined the ski area's relationship with the mountain. It ceased to be a 'laissez faire' ski area. Prior to snowmaking, Snowbowl opened when there was adequate snow, and closed when there was too little. And then there would be the matter of spreading large volumes of reclaimed water on the mountain.

People fought hard against snowmaking at Snowbowl and for good reason. Effluent or reclaimed water is treated municipal wastewater. Wherever Flagstaff's A+ class reclaimed water is used on the city's playing fields, parks and lawns, there is always signage advising against either drinking or bathing in it. The risk from exposure to it is underscored by the research of Dr. Cathy Propper, NAU biologist. Dr. Propper's research evaluated the impact of Flagstaff's A+ reclaimed water on the development of two species of amphibians. Her studies demonstrated that exposure to reclaimed water during their development significantly disrupted normal sexual development and thyroid function (Searcy et al. 2012, Wolff et al. 2015). Similar results have been found in studies of fish exposed to effluent and many species exposed to certain pesticides. Many of these endocrine-disrupting compounds are potent in remarkably low concentrations—even at parts per billion.

Despite these findings, Coconino National Forest, as a result of the 2005 Final Environmental Impact Statement (FEIS), stated that "the State of Arizona allows Class A and A+ reclaimed water for direct reuse in snowmaking. As stated in the FEIS, the reclaimed water (Class A+) which the City of Flagstaff provides to the ski resort meets all pertinent state and federal water quality standards" (Brady Smith, Coconino National Forest, personal communication, 2016).

Snowmaking won. In 2012, the 9th Circuit Court of Appeals ruled against the Save the Peaks Coalition, the group representing the Native American tribes, local citizens and environmentalists opposed to snowmaking. I attended a number of public meetings about this proposal in Flagstaff, and nobody put it together like the Native American tribes. They spoke so eloquently about the need for clean water and clean land, and that we cannot drink money. There were additional lawsuits, and a number of trials, sometimes with environmental challenges winning temporarily, but ultimately being reversed. The Supreme Court declined to hear the case, which in effect upheld the 9th Circuit's ruling.

Snowbowl signed a 20-year contract with the City of Flagstaff to purchase reclaimed water for snowmaking, starting in 2012. Basically, the ski area can purchase and make snow with up to 1.5 million gallons (5700 m^3) of effluent each day, between November and February. The City has permitted Snowbowl a total use of 178,000,000 gallons (674,000 m^3) of effluent on its slopes each winter.

There has been no monitoring of the impact of reclaimed water on the Peaks. Catherine Propper's General Statement for Wastewater for the Peaks states, "we do not know whether spreading the reclaimed wastewater effluent (note: effluent is different from sewage in that it has been treated) over ski slopes will spread the contamination farther and contaminate more of the surface waters and ground water, or whether exposure to freezing, thawing and UV along with soil microbial processes will help to eliminate the compounds; therefore, because the studies haven't been done, as a scientist, it is difficult to say what the impact of reclaimed water use will be on the slopes of the San Francisco Peaks" (Propper, personal communication, 2017).

As the 2005 FEIS states, there are a number of cattle and wildlife tanks, springs and wells downslope of Snowbowl, and the impact of effluent-laden snowmaking runoff on those water sources is unknown. Monitoring the quality of water emanating from those sites would reveal a great deal about the consequences of snowmaking with effluent on the Peaks. Wouldn't those be the perfect places to determine the extent of reclaimed water: places downslope of Snowbowl, where soils, plants, wild animals and cattle, water, school kids visiting Camp Colton and The Nature Conservancy's Hart Prairie Preserve, and homeowners might be impacted? Sampling in the Inner Basin would also be in order, considering the possibility that westerly winds may drive snow made from reclaimed water over the rim and into the Basin. It would be good to know, and it is totally doable. I mean, why not? A study of impacts on plants and animals would help, too.

Certainly there is another perspective on this. As Propper has implied, we live in a world utterly awash with human-made chemicals, many of which go into creating all of the things that we use. We are exposed to thousands of chemicals created for use in agriculture, industry, pharmacology and personal care products. While Propper has some reservations about the Peaks and snowmaking with effluent, I think she also sees it as a relatively minor event in the global context of chemical pollution. I can't go there, particularly in light of the fact that water in the Inner Basin is regarded as some of the cleanest in the nation (see above). So, yes, we are polluting much of the world, but do we have to pollute all of it?

Lawsuits are ongoing to stop snowmaking at Snowbowl.

White Vulcan Mine

In 1953, claims were filed on 320 acres (130 ha) of the eastern flank of the Peaks, and the White Vulcan Mine began mining pumice in 1968. In 1985, the U.S. Forest Service tried to render the claim null and void, but mining continued with a new pumice market, that of making stone-washed jeans. In the 1990s, Arizona Tufflite, Inc. sought to expand the mine, and in response the U.S. Forest Service filed an application to temporarily withdraw more than 74,000 acres (30,000 ha) around the Peaks from consideration for mining. Interior Secretary Bruce Babbitt joined opposition to the mining efforts in 1999, leading to closure of the mine along with a guarantee that the site would be fully restored within 5 years. In 2000, U.S. Interior and Agriculture departments actually paid Tufflite one million dollars to cease pumice mining at White Vulcan Mine. As part of the agreement, Tufflite relinquished all 49 of its mining claims on the Peaks, totalling 8,000 acres (3,240 ha), and the deal was completed by 2010. In 2010, volunteers restored some of the slopes of the mined area with plantings.

There are other mining projects, such as rock pit mining, slated for the north side of the Peaks.

Very minor deposits of gold have been found on the Peaks. Sand and gravel have been mined, and used in local building projects, from sources in lowlands surrounding the mountain, including Fort Valley, Kendrick Park and Black Bill Park. The Sinagua Formation was avoided by the mining industry due to a high degree of consolidation, and large boulders.

Protection

In 1902, President Theodore Roosevelt designated the Peaks and surrounding area as part of the San Francisco Mountains Forest Preserve, which was a critical, early effort in protecting the Peaks from overlogging and overgrazing (see John Leiberg above). Again, in the 1970s, private land at the base of the mountain was acquired by Coconino National Forest to protect it from future development.

In 1984, the Kachina Peaks Wilderness Area was designated, meaning that the area above 8,500 ft (2,600 m) on average—more than 18,000 acres (7,280 ha)—acquired additional protection. This act, established under President Jimmy Carter, prohibits logging, road building, mining, mechanized vehicles, etc., though grazing leases and mining claims are sometimes 'grandfathered' in, meaning that these agreements had been in place prior to the wilderness designation and still stand. That is the case on the Peaks, and the current lease-holders for the Peaks allotment—the Navajo Nation—have not made an effort to graze the mountain. And here's hoping that remains the case. Hunting is allowed, though a friend suggested that this high elevation wilderness area provides refuge for the burgeoning elk population.

Speaking of grazing, Coconino National Forest Range Specialist Gary Hase says that grazing of cattle and sheep around the Peaks is less intensive than it used to be. There is no longer year-round grazing, and pasture areas are typically grazed for no more than 30 to 60 days.

— SECTION 7 —

Sky Island Time and the Dry Days Ahead

The Peaks lives! Such a story it tells about mountain life in the Southwest, along the Western Cordillera. This story reflects an incredibly dynamic time during which this mountain came to be, and it ain't over yet. Endless change. I love the stories of plants from the Arctic living at the top of this mountain, and that the American Pipit flies in from the Rockies to breed in the tundra here. I love that limber pine and southwestern white pine meet here from the north and the south and hybridize. And that aspen stands are the loveliest of places and so prominent all around the mountain, reflecting a history of massive fire. That there are the coolest little communities of tundra plants hanging on wherever they can all the way to the top of the Peaks. That all these species of plants and animals have come together on this beautiful and remote island. And that this mountain receives hundreds and hundreds of millions of gallons of water from the sky each year.

Now climate change is upon us, upon this mountain, and it's going to be something. It is March, 2017, I am wearing flip flops and shorts while writing this book, here at 7,000 feet, near the base of the Peaks. It feels different, it feels wrong. Average annual temperatures in Flagstaff increased by 2 °F between 1971 and 2013 (Fig. 5.1; Hereford 2014). The second half of the 20th century was the warmest interval during the last 1,425 years, as measured with tree rings from bristlecone pines on the Peaks (Salzer & Kipfmueller 2005). Most local data sets also show a downturn in precipitation in all forms starting in 1996 (Fig. 5.2; Hereford 2014). They predict that drought will continue through the 21st century, exacerbated by increasing temperatures. In the Southwest, studies predict 'dustbowl-like' megadrought conditions lasting decades to a century (Cook et al. 2015, Fawcett et al. 2011, Seager et al. 2007). A decrease in winter precipitation and an increase in evaporation rates will fuel drought, and challenge both natural and human populations (Cook et al. 2015).

With climate change comes more variable and extreme conditions. I am struck by how we are seeing late frosts wreaking absolute havoc on plants. A hard frost in June of 1996 set off aspen decline on the Peaks and elsewhere in the Southwest, and that played out for a decade. A recent late spring frost in Flagstaff nailed even one of the last plants to leaf out each spring, Gambel oak. Whole canopies of its early summer leaves were frozen and turned brown. The early warmer temperatures are going to throw a wrench into the world's ecology. The 70 °F (21 °C) temperatures in February here led things to leaf out too early to accommodate spring freezes. It's hard to imagine how plants and animals can adapt to such increasingly variable conditions. USFS plant geneticist Gerald Rehfeldt suggests that if damaging frosts are uncommon, plants can tolerate the odd hit, but "if frosts increase in frequency and intensity, with wavy jet streams causing a switch from arctic to subtropical air masses overnight, well, vegetation as we know it will be challenged" (Gerald Rehfeldt, personal communication 2017).

Adams and Kolb (2005) found that all forest trees on the Peaks were negatively affected by drought in the 1950s to the 2000s, and especially at lower elevations. Clark Schaack (1983) even found that tundra-lover Whipple's penstemon didn't bloom during a dry year, and then did bloom the next year, with more precipitation, with snowfields melting into early June. This really surprised me, that water budgets can be tight at 12,000 ft (3,650 m), too. There is some evidence of climbing forests on the Peaks, with *Abies* occurring 650 feet higher and *Pinus flexilis* occurring 500 feet higher in elevation between 1876 and 2000 (Cocke et al. 2005).

Being a sky island in the Southwest in a time of climate warming is a big deal, since warming will enhance isolation and loss. In thinking about the Peaks as a sky island, I had been considering only the the tundra species and their distances from related populations, as they are already so far apart. The species that live above treeline on the Peaks—well, they will have nowhere else to go on this mountain, quite an image there. With such warming forecasted, many of the species species in the lower spruce-fir and mixed conifer zones may also become isolated from related populations.

While sky islands are a global phenomenon, their significance was pointed out early on for the many Madrean Sky Islands in southern Arizona and northern Mexico (SkyIslandsAlliance.org). Sky islands are isolated mountains surrounded by radically different lowland environments. The Sky Islands Alliance has been studying the Madrean Islands for a long time. The islands support the northern extent of Madrean oak and pine woodlands. This region is regarded as a gap in the Western Cordillera's span from Alaska to southern Mexico, and has a unique flora on its islands. The islands are Basin and Range mountains isolated from one another by intervening desert basins and valleys, usually below 4,000 ft (1,220 m). The Alliance studies nearly 40 small mountains in that region, with an elevational range of about 5,600 to 9,500 ft (1,700 to 2,900 m). They appear to be separated from one another by about 20-30 miles (~32-48 km) in general. The areas of the islands extend from several mountains that are less than 25 acres (10 ha) to the Rincon Mountains, which are over 66,000 acres (26,700 ha). Some of the islands have a high diversity of plants and animals. The small sky island, Tumacácori

(325 acres, 132 ha), has over 300 species of plants and 129 species of birds associated with it! And these measures are from the early 2000s (2000-2003), which were drought years in the Southwest! Near Tucson, the Rincon Mountains (67,000 acres, 27,186 ha) support 302 plants, while the Tucson Mountains (240,000 acres, 97,120 ha) support 177 species of plants. These sky islands house a number of endemic species, perhaps due in part to their great age as well as isolation, having probably arisen between 6 and 35 Ma.

The Peaks—a sky island of the Colorado Plateau—covers 56,732 acres (23,000 ha), and reaches 12,633 ft (3,800 m). It holds its own in the biodiversity department by supporting at least 100 species of fungi, 567 plant taxa, 161 species of birds, 40 mammals, and a few reptiles and amphibians. Obviously no fish occur, for which we can thank the very leaky nature of this stratovolcano's structure. There are endemic fish in spring-fed streams near the Peaks, though!

The Peaks supports 3 endemic species of plants, though there may well be more, and hopefully this book will encourage more of that type of research. More than 30 plant species occur only on the Peaks within Arizona, and mostly in its upper zones, separated from gene flow with related populations by hundreds and hundreds of miles. They range from little annual tundra plants to bristlecone pine, and they somehow managed to find the Peaks during the Pleistocene. The notion of their migration to the Peaks will always fascinate me. They have probably been separated from related populations for 10,000 years.

While most high elevation plants on the Peaks represent range extensions of Rocky Mountain plants, many lower elevation species on the mountain hail from the south. Both limber pine and southwestern white pine are ultimately from Mexico and Central America. It would be wonderful to know how the assemblage of plants of the Peaks, as a sky island with its isolation and elevation, compares with the wash of species from south to north along the Western Cordillera, which to me means the western mountain highway. This is a project I hope will happen. In working on this book, I did notice that a number of plant species on the Peaks have distributions that span Arizona.

There is also the matter of plant life along the uplifted rim of the Colorado Plateau, which supports most forest species (Plate 3, Fig. 3.10)! There are many, many mountain species along this uplifted circle of mountains, with good expanses of species' populations over large distances. The Peaks and the Mogollon Rim are part of this circle of suitable habitat, this highway of higher elevation life, and a subset of the Cordillera.

Finally

The days ahead are going to be different, maybe for every creature on Earth. And you know it's going to be very interesting. I hope this book, with its timeline of endless action, will inspire and maybe provide hope. I just love that even the ancient conifers have been diverging wildly since the dry times started 10,000 years ago, getting better adapted quite quickly to conditions all along the Cordillera.

Drier times for the mountain will mean that we all need to approach the place a lot more gently. Those of us who live nearby, who get to see the Peaks translucent with early light and with late light—growing out of the sky each day—we do love this mountain. Someone that I interviewed for this book, who is a particularly avid Peaks hiker, said that he, volunteering with Forest Service crews, has hauled down tons and tons of garbage. Wow! No more garbage, please, only love.

— APPENDIX I —

Plants of the San Francisco Peaks

Appendix I. The plant list for the San Francisco Peaks is organized by taxonomy, and by life zone, see Table 3.2. Max Licher and Gwen Waring organized this list, working from lists provided by William Moir, the Deaver Herbarium, NAU, and SEINET. Additional support was provided by Kirsten Phillips, Museum of Northern Arizona, and George Ferguson, University of Arizona.

Note that numbers of Liverworts and Mosses are included in the total species count of 567 species, but not in habitat counts, in this edition.

Family	Scientific Name	Common Name
LIVERWORTS		
LOPHOZIACEAE	*Lophozia longiflora* (=*L. ventricosa ssp. longiflora; historic synonym: L. porphyroleuca*)	None – leafy liverwort
MARCHANTIACEAE	*Marchantia polymorpha*	
MOSSES		
AMBLYSTEGIACEAE	*Amblystegium serpens*	
ANDREAEACEAE	*Andreaea rupestris*	
ANOMODONTACEAE	*Anomodon rostratus*	
BRATRAMIACEAE	*Philonotis fontana*	
BRACHYTHECIACEAE	*Brachytheciastrum collinum*	None- pleurocarp
	Brachythecium rivulare	
BRYACEAE	*Bryum argenteum*	
	Ptychostomum creberrimum	
	Ptychostomum pallescens	
	Ptychostomum pseudotriquetrum	
	Rosulabryum flaccidum	
DICRANACEAE	*Dicranum scoparium*	None- acrocarp
DITRICACEAE	*Ceratodon purpureus*	None- acrocarp
GRIMMIACEAE	*Grimmia alpestris* (historic synonym: *Grimmia donniana alpestris*)	Grimmia dry rock moss
	Grimmia arizonae	Arizona dry rock moss
	Jaffueliobryum raui (historic synonym: *Grimmia raui*)	None - acrocarp
HYPNACEAE	*Hypnum cupressiforme*	
	Hypnum revolutum	Revolute Hypnum moss
LESKEACEAE	*Pseudoleskeella tectorum* (historic synonym: *Leskea crytophylla)*	None - pleurocarp
	Pseudoleskea incurvata (historic synonym: *P. oligoclada*)	None - pleurocarp
MEESIACEAE	*Leptobryum pyriforme*	
MIELICHHOFERIACEAE	*Pohlia nutans*	
ORTHOTRICHACEAE	*Orthotrichum spp.*	None - acrocarp

Family	Scientific Name	Common Name	Ponderosa pine	Mixed conifer	Spruce-fir	Tundra	Nonnative
MNIACEAE	*Mnium arizonicum*						
POLYTRICHACEAE	*Polytrichum juniperinum*						
POTTIACEAE	*Syntrichia norvegica*						
	Syntrichia ruralis	Twisted moss					
	Weissia controversa (historic synonym: *W. viridula*)	Puzzle moss					
RHABDOWEISIACEAE	*Dicranoweisia cirrata*	None- acrocarp					
	Dicranoweisia crispula	None- acrocarp					
HORSETAILS							
EQUISETACEAE	*Equisetum laevigatum*	smooth horsetail	1	1			
FERNS							
ASPLENIACEAE	*Asplenium septentroinale*	forked spleenwort	1	1			
DENNSTAEDTIACEAE	*Pteridium aquilinum*	western brackenfern	1	1			
DRYOPTERIDACEAE	*Dryopteris filix-mas*	male fern	1	1			
MARSILEACEAE	*Marsilea mollis*	Chihuahuan waterclover	1				
OPHIOGLOSSACEAE	*Botrychium crenulatum*	scalloped moonwort			1		
	Botrychium echo	reflected grapefern			1	1	
	Botrychium hesperium	western moonwort			1		
	Botrychium lanceolatum	lanceleaf grapefern		1	1		
	Botrychium lunaria	common moonwort			1	1	
	Botrychium minganense	Mingan moonwort			1		
POLYPODIACEAE	*Polypodium hesperium*	western polypody	1	1			
WOODSIACEAE	*Cystopteris fragilis*	brittle bladderfern			1		
	Cystopteris reevesiana	Reeves' bladderfern		1	1	1	
	Woodsia neomexicana	New Mexico cliff fern	1	1			
	Woodsia oregana	Oregon cliff fern	1	1			
	Woodsia oregana subsp. *cathcartiana*	Oregon cliff fern	1				
CONIFERS							
CUPRESSACEAE	*Juniperus communis* var. *depressa*	common juniper	1	1	1	1	
	Juniperus osteosperma	Utah juniper	1	1			
PINACEAE	*Abies concolor*	white fir		1	1		
	Abies lasiocarpa var. *arizonica*	corkbark fir		1	1	1	
	Picea engelmannii	Engelmann spruce	1	1	1		
	Picea pungens	blue spruce	1	1	1		
	Pinus aristata ●	bristlecone pine	1	1	1	1	
	Pinus edulis	twoneedle pinyon	1	1			
	Pinus flexilis	limber pine	1	1	1		
	Pinus ponderosa	ponderosa pine	1	1			
	Pinus strobiformis	southwestern white pine	1	1	1		
	Pseudotsuga menziesii	Douglas-fir	1	1	1		
FLOWERING PLANTS							
ADOXACEAE	*Sambucus nigra* subsp. *caerulea*	blue elderberry	1	1	1		
	Sambucus racemosa var. *racemosa*	red elderberry		1	1		
ALISMATACEAE	*Alisma triviale*	northern water plantain		1	1		
AMARANTHACEAE	*Amaranthus powellii*	Powell's amaranth	1	1	1		
	Chenopodium album	lambsquarters	1				x
	Chenopodium berlandieri	pitseed goosefoot	1	1	1		
	Chenopodium capitatum	blite goosefoot		1	1		
	Chenopodium fremontii	Fremont's goosefoot	1	1			
	Chenopodium pratericola	desert goosefoot	1	1			
	Dysphania graveolens	fetid goosefoot	1	1			
	Kochia scoparia	mock cypress	1	1	1		x
	Salsola tragus	prickly Russian thistle	1	1			x
AMARYLLIDACEAE	*Allium bisceptrum* var. *palmeri*	aspen onion		1			

Family	Scientific Name	Common Name	Ponderosa pine	Mixed conifer	Spruce-fir	Tundra	Nonnative
	Allium geyeri	Geyer's onion	1	1			
APIACEAE	*Osmorhiza depauperata*	bluntseed sweetroot	1	1			
	Perideridia parishii	Parish's yampah	1	1			
	Pseudocymopterus montanus	alpine false springparsley	1	1	1	1	
APOCYNACEAE	*Apocynum androsaemifolium*	spreading dogbane	1	1	1		
	Apocynum x floribundum	dogbane	1	1			
	Asclepias asperula	spider milkweed	1				
	Asclepias speciosa	showy milkweed	1	1			
	Asclepias subverticillata	horsetail milkweed	1				
ASPARAGACEAE	*Echeandia flavescens*	Torrey's craglily	1	1			
	Maianthemum racemosum	feathery false lily of the vally		1			
	Maianthemum stellatum	starry false lily of the vally		1	1		
ASTERACEAE	*Achillea millefolium* var. *occidentalis*	western yarrow	1	1	1		
	Ageratina herbacea	fragrant snakeroot	1	1			
	Agoseris aurantiaca	orange agoseris	1	1	1		
	Agoseris glauca	pale agoseris	1	1	1		
	Amauriopsis dissecta	ragleaf bahia	1				
	Ambrosia tomentosa	burragweed	1	1			
	Antennaria marginata	whitemargin pussytoes	1	1	1		
	Antennaria media	Rocky Mountain pussytoes			1	1	
	Antennaria microphylla	littleleaf pussytoes	1	1		1	
	Antennaria parvifolia	small-leaf pussytoes	1	1	1		
	Antennaria rosea	rosy pussytoes	1	1	1	1	
	Antennaria rosulata	Kaibab pussytoes	1	1		1	
	Antennaria umbrinella	umber pussytoes			1	1	
	Arnica cordifolia	heartleaf arnica	1	1	1		
	Artemisia carruthii	Carruth's sagewort	1	1	1		
	Artemisia dracunculus	wild tarragon	1	1	1		
	Artemisia ludoviciana subsp. *albula*	white sagebrush		1	1		
	Brickellia californica	brickell-bush	1				
	Brickellia eupatorioides var. *chlorolepis*	false boneset	1	1	1		
	Brickellia grandiflora	tasselflower brickellbush	1	1	1		
	Centaurea diffusa	diffuse knapweed	1				x
	Cirsium arizonicum	Arizona thistle	1	1	1		
	Cirsium parryi	Parry's thistle		1	1		
	Cirsium scariosum	elk thistle	1	1			
	Cirsium vulgare	bull thistle	1				x
	Cirsium wheeleri	Wheeler's thistle	1	1	1		
	Conyza canadensis	Canadian horseweed	1				
	Coreopsis tinctoria	golden tickseed	1	1			
	Cosmos parviflorus	southwestern cosmos	1				
	Ericameria nauseosa	rubber rabbitbrush	1				
	Erigeron compositus ●	cutleaf daisy			1	1	
	Erigeron divergens	spreading fleabane	1	1	1		
	Erigeron eximius	sprucefir fleabane		1	1	1	
	Erigeron flagellaris	trailing fleabane	1	1	1		
	Erigeron formosissimus	beautiful fleabane	1	1	1		
	Erigeron neomexicanus	New Mexico fleabane	1	1			
	Erigeron oreophilus	chaparral fleabane	1				
	Erigeron simplex ●	onestem fleabane			1	1	
	Erigeron speciosus	aspen fleabane	1	1	1		
	Erigeron tracyi	running fleabane	1				
	Erigeron vreelandii	Vreelands erigeron	1	1			
	Gnaphalium exilifolium	slender cudweed	1	1			
	Grindelia squarrosa	curlycup gumweed	1				
	Gutierrezia sarothrae	broom snakeweed	1				
	Helianthella quinquenervis	fivenerve helianthella	1	1	1		
	Helianthus annuus	common sunflower	1				

Family	Scientific Name	Common Name	Ponderosa pine	Mixed conifer	Spruce-fir	Tundra	Nonnative
	Helianthus petiolaris	prairie sunflower	1				
	Heliomeris multiflora var. *multiflora*	showy goldeneye	1	1			
	Heliomeris multiflora var. *nevadensis*	Nevada goldeneye	1	1			
	Heterosperma pinnatum	wingpetal	1				
	Heterotheca villosa	hairy false goldenaster	1	1			
	Hieracium fendleri	yellow hawkweed	1	1	1		
	Hymenopappus mexicanus	Mexican woollywhite	1	1			
	Hymenoxys bigelovii	Bigelow's rubberweed	1				
	Hymenoxys helenioides	Intermountain rubberweed	1				
	Hymenoxys hoopesii	sneezeweed	1	1	1	1	
	Hymenoxys richardsonii	Colorado rubberweed	1	1			
	Lactuca serriola	prickly lettuce	1				x
	Laennecia coulteri	Coulter's horseweed			1		
	Laennecia schiedeana	pineland marshtail	1	1	1		
	Machaeranthera canescens	hoary tansyaster	1	1			
	Machaeranthera tanacetifolia	tanseyleaf tansyaster	1				
	Oreochrysum parryi	Parry's goldenrod		1	1		
	Packera franciscana ● *	San Francisco Peaks ragwort			1	1	
	Packera multilobata	lobeleaf groundsel		1	1		
	Packera neomexicana var. *mutabilis*	New Mexico groundsel	1	1	1		
	Packera werneriifolia	hoary groundsel			1		
	Pericome caudata	mountain tail-leaf		1	1		
	Pseudognaphalium macounii	Macoun's cudweed	1	1	1		
	Pyrrocoma crocea	curlyhead goldenweed	1	1			
	Rudbeckia laciniata	cutleaf coneflower	1	1			
	Senecio bigelovii	nodding ragwort		1	1		
	Senecio eremophilus var. *macdougalii*	MacDougal's ragwort	1	1	1		
	Senecio spartioides	broom groundsel	1	1			
	Senecio wootonii	Wooton's ragwort	1	1	1		
	Solidago canadensis var. *scabra*	Canada goldenrod	1				
	Solidago missouriensis	Missouri goldenrod	1				
	Solidago multiradiata	Rocky Mountain goldenrod			1	1	
	Solidago nana	baby goldenrod	1	1			
	Solidago simplex var. *simplex*	Mt. Albert goldenrod	1	1	1		
	Solidago velutina	threenerve goldenrod	1	1			
	Solidago wrightii	Wright's goldenrod	1	1			
	Stephanomeria minor	lesser wirelettuce	1				
	Stephanomeria thurberi	Thurber's wirelettuce	1				
	Symphyotrichum ascendens	western aster	1				
	Symphyotrichum falcatum var. *commutatum*	white prairie aster	1	1			
	Symphyotrichum foliaceum	alpine leafybract aster	1	1			
	Taraxacum laevigatum	rock dandelion				1	x
	Taraxacum lyratum	black-seeded dandelion				1	
	Taraxacum officinale	common dandelion	1	1	1	1	x
	Tetradymia canescens	spineless horsebrush	1				
	Tragopogon dubius	yellow salsify	1				x
	Tragopogon pratensis	meadow salsify	1	1			x
	Wyethia arizonica	Arizona mule-ears	1				
	Xanthisma gracile	slender goldenweed	1				
BERBERIDACEAE	*Berberis repens*	creeping barberry	1				
BORAGINACEAE	*Cryptantha cinerea* var. *cinerea*	James' cryptantha	1	1			
	Cryptantha fendleri	sanddune cryptantha	1				
	Cryptantha setosissima	bristly cryptantha	1	1	1		
	Cynoglossum officinale	gypsyflower		1			x
	Hackelia floribunda	manyflower stickseed	1	1			
	Lappula occidentalis	flatspine stickseed	1	1	1		
	Lappula squarrosa	European stickseed		1			x

Family	Scientific Name	Common Name	Ponderosa pine	Mixed conifer	Spruce-fir	Tundra	Nonnative
	Lithospermum cobrense	smooththroat stoneseed	1				
	Lithospermum incisum	narrowleaf stoneseed	1				
	Lithospermum multiflorum	manyflowered stoneseed	1	1	1		
	Macromeria viridiflora	giant-trumpets	1	1			
	Mertensia franciscana	Franciscan bluebells		1	1	1	
	Nama dichotomum	wishbone fiddleleaf	1	1			
	Phacelia heterophylla	varileaf phacelia	1				
	Plagiobothrys scouleri	Scouler's popcornflower	1	1			
BRASSICACEAE	*Barbarea orthoceras*	American yellowrocket	1				
	Boechera fendleri	Fendler's rockcress	1				
	Capsella bursa-pastoris	shepherd's purse	1				x
	Cardamine cordifolia	heartleaf bittercress		1	1		
	Descurainia incisa subsp. *paysonii*	Payson's tansymustard	1				
	Descurainia obtusa	blunt tansymustard	1				
	Draba aurea	golden draba			1	1	
	Draba crassifolia ●	snowbed draba			1	1	
	Draba helleriana	Heller's draba			1		
	Hesperidanthus linearifolius	slimleaf plainsmustard	1	1			
	Lepidium densiflorum	common pepperweed	1				
	Pennellia longifolia	longleaf mock thelypody	1	1	1		
	Rorippa sphaerocarpa	roundfruit yellowcress	1	1			
	Thlaspi montanum var. *fendleri*	Fendler's pennycress	1	1	1	1	
CAMPANULACEAE	*Campanula parryi*	Parry's bellflower	1	1	1		
	Campanula rotundifolia	bluebell bellflower	1	1			
	Porterella carnosula	fleshy porterella	1				
CANNABACEAE	*Cannabis sativa*	marijuana	1	1	1		x
CAPRIFOLIACEAE	*Dipsacus fullonum*	Fuller's teasel	1				x
	Lonicera arizonica	Arizona honeysuckle	1	1			
	Lonicera involucrata	twinberry honeysuckle	1	1	1	1	
	Symphoricarpos oreophilus	mountain snowberry	1	1	1		
	Symphoricarpos rotundifolius var. *parishii*	Parish's snowberry	1	1	1		
	Valeriana acutiloba var. *acutiloba*	sharpleaf valerian		1	1	1	
	Valeriana arizonica	Arizona valerian		1			
	Valeriana edulis	tobacco root	1	1	1		
CARYOPHYLLACEAE	*Arenaria fendleri*	Tweedy's sandwort	1	1	1	1	
	Arenaria fendleri var. *tweedyi*	Tweedy's sandwort	1	1	1	1	
	Arenaria lanuginosa var. *saxosa*	spreading sandwort	1	1	1	1	
	Cerastium beeringianum ●	Bering chickweed			1	1	
	Cerastium brachypodum	shortstalk chickweed				1	
	Cerastium fontanum	common mouse-ear chickweed		1			x
	Drymaria effusa var. *depressa*	pinewoods drymary	1	1			
	Drymaria molluginea	slimleaf drymary	1				
	Minuartia obtusiloba ●	twinflower sandwort			1	1	
	Minuartia rubella ●	beautiful sandwort			1	1	
	Minuartia stricta ●	sandwort			1	1	
	Sagina saginoides	arctic pearlwort		1	1	1	
	Silene acaulis ●	moss campion			1	1	
	Silene drummondii var. *drummondii*	Drummond's campion	1				
	Silene latifolia subsp. *alba*	Eurasian campion	1				
	Silene scouleri	simple campion	1	1	1		
	Stellaria umbellata	umbrella starwort		1	1	1	
CLEOMACEAE	*Peritoma serrulata*	Rocky Mountain beeplant	1				
COMMELINACEAE	*Commelina dianthifolia*	birdbill dayflower	1	1	1		
CONVOLVULACEAE	*Convolvulus arvensis*	field bindweed	1	1			x
CORNACEAE	*Cornus sericea* subsp. *sericea*	red osier dogwood	1	1			
CRASSULACEAE	*Rhodiola rhodantha*	redpod stonecrop			1	1	
CYPERACEAE	*Carex albonigra* ●	black and white sedge				1	
	Carex athrostachya	slenderbeak sedge		1	1		

Family	Scientific Name	Common Name	Ponderosa pine	Mixed conifer	Spruce-fir	Tundra	Nonnative
	Carex bella	southwestern showy sedge			1	1	
	Carex chalciolepis ●	Holm sedge				1	
	Carex deflexa var. *boottii* ●	northern sedge				1	
	Carex ebenea	ebony sedge			1	1	
	Carex elynoides ●	blackroot sedge				1	
	Carex geophila	White Mountain sedge	1	1			
	Carex haydeniana ●	cloud sedge				1	
	Carex microptera	smallwing sedge	1	1			
	Carex nebrascensis	Nebraska sedge	1	1			
	Carex occidentalis	western sedge	1	1	1		
	Carex oreocharis	grassyslope sedge	1	1			
	Carex pellita	woolly sedge	1	1			
	Carex petasata	Liddon sedge	1	1			
	Carex rossii	Ross' sedge		1	1		
	Carex siccata	dryspike sedge	1	1	1	1	
	Carex subfusca	brown sedge	1	1			
	Carex wootonii	Wooton's sedge		1	1		
	Cyperus fendlerianus	Fendler's flatsedge	1				
	Cyperus squarrosus	bearded flatsedge		1			
	Eleocharis acicularis	needle spikerush	1	1			
	Eleocharis palustris	common spikerush	1	1			
	Eleocharis quinqueflora	fewflower spikerush		1			
ELATINACEAE	*Elatine brachysperma*	shortseed waterwort		1			
	Elatine californica	California waterwort		1			
ERICACEAE	*Chimaphila umbellata* subsp. *acuta*	pipsissewa	1				
	Moneses uniflora	single delight			1		
	Monotropa hypopitys	pinesap	1	1			
	Orthilia secunda	sidebells wintergreen		1	1		
	Pterospora andromedea	woodland pinedrops	1	1			
	Pyrola chlorantha	greenflowered wintergreen		1	1		
	Pyrola minor	snowline wintergreen			1		
EUPHORBIACEAE	*Chamaesyce serpyllifolia*	thymeleaf sandmat	1				
	Euphorbia brachycera	horned spurge		1	1		
	Euphorbia lurida	woodland spurge	1				
FABACEAE	*Astragalus hallii*	Hall's milkvetch	1				
	Astragalus humistratus var. *hosackiae*	ground-cover milkvetch	1				
	Astragalus humistratus var. *humivagans*	ground-cover milkvetch	1				
	Astragalus rusbyi	Rusby's milkvetch	1	1			
	Hedysarum boreale	northern sweetvetch			1		
	Lathyrus graminifolius	Grassleaf peavine	1				
	Lathyrus lanszwertii var. *leucanthus*	Nevada pea	1	1	1		
	Lotus corniculatus	ground honeysuckle		1			x
	Lotus wrightii	Wright's deervetch	1	1			
	Lupinus argenteus	silvery lupine	1	1	1		
	Lupinus kingii	King's lupine	1	1			
	Medicago lupulina	black medick	1	1	1		x
	Medicago sativa	alfalfa	1	1			x
	Melilotus officinalis	yellow sweetclover	1	1	1		x
	Oxytropis lambertii	purple locoweed	1	1	1		
	Phaseolus angustissimus	slimleaf bean	1				
	Phaseolus grayanus	Gray's bean	1	1	1		
	Robinia neomexicana	New Mexico locust		1			
	Thermopsis montana	mountain goldenbanner	1	1			
	Trifolium hybridum	alsike clover	1	1	1		x
	Trifolium longipes subsp. *pygmaeum*	pygmy clover	1	1			
	Trifolium wormskioldii	cows clover	1	1			
	Vicia americana	American vetch	1	1	1		
	Vicia pulchella	sweetclover vetch	1	1			

Family	Scientific Name	Common Name	Ponderosa pine	Mixed conifer	Spruce-fir	Tundra	Nonnative
	Vicia villosa	winter vetch	1	1	1		x
FAGACEAE	*Quercus gambelii*	Gambel oak	1	1			
GENTIANACEAE	*Frasera speciosa*	elkweed		1	1		
	Gentiana affinis	pleated gentian	1	1	1		
	Gentiana fremontii	moss gentian		1			
	Gentianella amarella subsp. *acuta*	autumn dwarf gentian			1	1	
	Gentianella amarella subsp. *heterosepala*	autumn dwarf gentian			1	1	
	Gentianella tenella ●	Dane's dwarf gentian			1	1	
	Gentianopsis barbellata ●	perennial fringed gentian			1	1	
GERANIACEAE	*Erodium cicutarium*	redstem storksbill	1				x
	Geranium caespitosum	pineywoods geranium	1	1	1		
	Geranium richardsonii	Richardson's geranium	1	1	1		
GROSSULARIACEAE	*Ribes cereum*	wax currant	1	1			
	Ribes montigenum	gooseberry currant	1	1	1	1	
	Ribes pinetorum	orange gooseberry		1	1		
HALORAGACEAE	*Myriophyllum sibiricum*	shortspike watermilfoil	1				
HYDRANGEACEAE	*Philadelphus microphyllus*	littleleaf mock orange		1			
HYDROCHARITACEAE	*Elodea bifoliata*	twoleaf waterweed		1			
HYPERICACEAE	*Hypericum scouleri* subsp. *scouleri*	Scouler's St. Johnswort	1	1			
IRIDACEAE	*Iris missouriensis*	Rocky Mountain iris	1	1	1		
	Sisyrinchium demissum	stiff blue-eyed grass	1	1	1		
	Sisyrinchium longipes	timberland blue-eyed grass		1			
JUNCACEAE	*Juncus balticus* subsp. *ater*	rush	1	1			
	Juncus bufonius	toad rush	1	1			
	Juncus confusus	Colorado rush		1			
	Juncus drummondii ●	Drummond's rush				1	
	Juncus saximontanus	Rocky Mountain rush	1	1			
	Juncus longistylis	longstyle rush	1	1			
	Juncus nevadensis	Sierra rush		1			
	Juncus interior	poverty rush	1				
	Luzula parviflora	smallflowered woodrush		1	1		
	Luzula spicata ●	spiked woodrush			1	1	
LAMIACEAE	*Agastache pallidiflora*	Bill Williams Mountain giant hyssop		1			
	Dracocephalum parviflorum	American dragonhead	1				
	Hedeoma hyssopifolia	mock pennyroyal		1			
	Marrubium vulgare	horehound	1				x
	Mentha arvensis	wild mint		1			
	Monarda fistulosa var. *menthifolia*	wild bergamot	1	1			
	Monardella odoratissima	bee balm	1	1	1		
	Prunella vulgaris	common selfheal	1	1			
LENTIBULARIACEAE	*Utricularia macrorhiza*	common bladderwort		1			
LILIACEAE	*Calochortus ambiguus*	doubting mariposa lily	1	1	1		
	Prosartes trachycarpa	wartberry fairybell		1	1		
LINACEAE	*Linum aristatum*	bristle flax		1			
	Linum australe	southern flax	1				
	Linum lewisii	Lewis flax	1	1	1		
LOASACEAE	*Mentzelia rusbyi*	Rusby's blazingstar	1	1			
MALVACEAE	*Sidalcea neomexicana*	salt spring checkerbloom	1	1	1		
	Sphaeralcea fendleri subsp. *fendleri*	Fendler's globemallow	1				
MELANTHIACEAE	*Zygadenus elegans*	mountain deathcamas		1	1		
MONTIACEAE	*Montia chamissoi*	water minerslettuce	1				
NYCTAGINACEAE	*Mirabilis albida*	white four o'clock			1		
	Mirabilis linearis var. *decipiens*	narrowleaf four o'clock	1	1			
ONAGRACEAE	*Chamerion angustifolium*	fireweed		1	1	1	
	Epilobium brachycarpum	tall annual willowherb	1				
	Epilobium ciliatum	fringed willowherb	1	1	1		
	Epilobium saximontanum	Rocky Mountain willowherb		1	1	1	
	Oenothera caespitosa	tufted evening primrose	1	1	1		

Family	Scientific Name	Common Name	Ponderosa pine	Mixed conifer	Spruce-fir	Tundra	Nonnative
	Oenothera coronopifolia	crownleaf evening primrose	1				
	Oenothera elata	Hooker's evening primrose	1				
	Oenothera flava	yellow evening primrose	1	1			
	Oenothera laciniata	cutleaf evening primrose	1	1			
	Oenothera neomexicana	New Mexico evening primrose	1	1			
	Oenothera pubescens	silky evening primrose	1				
ORCHIDACEAE	*Calypso bulbosa*	fairy slipper		1	1		
	Corallorhiza maculata	spotted coralroot	1	1	1		
	Corallorhiza wisteriana	spring coralroot		1	1		
	Goodyera oblongifolia	western rattlesnake plantain		1			
	Spiranthes romanzoffiana	hooded lady's tresses		1			
OROBANCHACEAE	*Castilleja austromontana*	Rincon Mtn Indian paintbrush	1	1	1		
	Castilleja integra	wholeleaf Indian paintbrush	1	1			
	Castilleja miniata	giant red Indian paintbrush	1	1	1		
	Orobanche fasciculata	clustered broomrape	1				
	Orthocarpus luteus	yellow owl's-clover	1	1	1		
	Pedicularis parryi	Parry's lousewort			1	1	
OXALIDACEAE	*Oxalis decaphylla*	tenleaf woodsorrel	1	1			
PAPAVERACEAE	*Corydalis aurea*	scrambled eggs	1				
PHRYMACEAE	*Mimulus guttatus*	seep monkeyflower	1	1	1		
	Mimulus primuloides	primrose monkeyflower		1			
	Mimulus tilingii ●	Tiling's monkeyflower			1		
PLANTAGINACEAE	*Besseya arizonica*	Arizona coraldrops	1	1	1		
	Callitriche palustris	vernal water-starwort	1	1			
	Linaria dalmatica	Dalmatian toadflax	1	1	1		x
	Penstemon barbatus	beardlip penstemon	1	1	1		
	Penstemon putus	Black River beardtongue	1	1			
	Penstemon virgatus	Wandbloom penstemon	1				
	Penstemon whippleanus	Whipple's penstemon			1	1	
	Plantago argyraea	saltmeadow plantain	1				
	Plantago major	common plantain	1				x
	Veronica peregrina subsp. *xalapensis*	hairy purslane speedwell	1	1			
	Veronica wormskjoldii	American alpine speedwell			1	1	
POACEAE	*Agropyron desertorum*	desert wheatgrass	1		1		x
	Agrostis exarata	spike bentgrass	1	1	1		
	Agrostis idahoensis	Idaho bentgrass			1	1	
	Agrostis scabra	rough bentgrass	1	1	1		
	Agrostis stolonifera	creeping bentgrass	1	1			x
	Alopecurus aequalis	shortawn foxtail	1	1			
	Alopecurus carolinianus	Carolina foxtail		1			x
	Alopecurus geniculatus	water foxtail	1	1			x
	Aristida arizonica	Arizona threeawn	1				
	Blepharoneuron tricholepis	pine dropseed	1	1	1		
	Bouteloua curtipendula	sideoats grama	1				
	Bouteloua gracilis	blue grama	1				
	Bouteloua simplex	matted grama	1				
	Bromus carinatus	California brome	1	1			x
	Bromus ciliatus	fringed brome		1			
	Bromus inermis	smooth brome	1	1	1		x
	Bromus lanatipes	woolly brome	1	1			
	Bromus marginatus	mountain brome	1	1			
	Bromus polyanthus	Great Basin brome	1				
	Bromus porteri	Porter brome	1	1	1	1	
	Bromus tectorum	cheatgrass	1	1			x
	Calamagrostis canadensis	bluejoint			1		
	Dactylis glomerata	orchardgrass	1	1	1		x
	Danthonia intermedia	timber oatgrass			1		
	Deschampsia caespitosa	MacKenzie's hairgrass		1			

Family	Scientific Name	Common Name	Ponderosa pine	Mixed conifer	Spruce-fir	Tundra	Nonnative
	Elymus elymoides	squirreltail	1				
	Elymus lanceolatus	streambank wheatgrass	1	1			
	Elymus scribneri	spreading wheatgrass			1	1	
	Elymus trachycaulus	slender wheatgrass	1	1	1	1	
	Elymus violaceus ●	Arctic wheatgrass				1	
	Eragrostis mexicana	Mexican lovegrass	1				
	Festuca arizonica	Arizona fescue	1	1	1		
	Festuca brachyphylla var. *coloradensis*	Colorado fescue			1	1	
	Festuca calligera	southwest fescue			1		
	Festuca earlei	Earle's fescue			1	1	
	Festuca minutiflora ●	smallflower fescue			1	1	
	Festuca ovina	sheep fescue	1	1	1		x
	Festuca rubra	red fescue		1	1		x
	Festuca saximontana	Rocky Mountain fescue			1		
	Festuca sororia	ravine fescue			1		
	Festuca trachyphylla	hard fescue		1	1		x
	Glyceria borealis	small floating mannagrass	1	1			
	Glyceria striata	fowl mannagrass		1			
	Hesperostipa comata	needle and thread	1				
	Hordeum brachyantherum	meadow barley	1	1			
	Hordeum jubatum	foxtail barley	1		1		
	Koeleria macrantha	prairie Junegrass	1	1	1		
	Lolium perenne	perennial ryegrass	1	1	1		x
	Lolium pratense	meadow ryegrass	1				x
	Muhlenbergia minutissima	annual muhly	1	1			
	Muhlenbergia montana	mountain muhly	1	1	1		
	Muhlenbergia ramulosa	green muhly	1	1			
	Muhlenbergia richardsonis	mat muhly	1				
	Muhlenbergia rigens	deergrass	1	1			
	Muhlenbergia virescens	screwleaf muhly	1	1			
	Muhlenbergia wrightii	spike muhly	1	1	1		
	Pascopyrum smithii	western wheatgrass	1	1			
	Phalaris arundinacea	reed canarygrass		1			
	Phleum alpinum	alpine timothy		1	1	1	
	Phleum pratense	timothy	1	1	1		x
	Piptochaetium pringlei	Pringle's speargrass	1				
	Poa compressa	Canada bluegrass	1	1			x
	Poa fendleriana	muttongrass	1	1	1	1	
	Poa glauca	glaucous bluegrass			1	1	
	Poa nervosa	Veiny Bluegrass	1	1			
	Poa palustris	fowl bluegrass		1			
	Poa pratensis	Kentucky bluegrass	1	1	1	1	x
	Poa reflexa	nodding bluegrass		1	1	1	
	Poa secunda	Sandberg bluegrass	1	1	1	1	
	Schizachyrium scoparium	little bluestem	1				
	Thinopyrum intermedium	intermediate wheatgrass	1	1	1		x
	Thinopyrum ponticum	tall wheatgrass	1	1			x
	Trisetum spicatum	spike trisetum		1	1	1	
POLEMONIACEAE	*Collomia linearis*	tiny trumpet	1	1			
	Ipomopsis aggregata	scarlet gilia	1	1	1		
	Ipomopsis arizonica	Arizona ipomopsis	1	1			
	Ipomopsis multiflora	manyflowered ipomopsis	1	1			
	Ipomopsis tenuituba subsp. *latiloba*	slendertube skyrocket	1	1			
	Navarretia intertexta subsp. *propinqua*	near navarretia	1	1	1		
	Polemonium pulcherrimum subsp. *delicatum*	Jacob's-ladder				1	
	Polemonium viscosum	sticky polemonium				1	
POLYGONACEAE	*Bistorta bistortoides*	American bistort	1				

Family	Scientific Name	Common Name	Ponderosa pine	Mixed conifer	Spruce-fir	Tundra	Nonnative
	Eriogonum cernuum	nodding buckwheat	1				
	Eriogonum polycladon	sorrel buckwheat	1				
	Eriogonum racemosum	redroot buckwheat	1	1			
	Eriogonum umbellatum var. *cognatum*	sulphur-flower buckwheat	1				
	Oxyria digyna ●	alpine mountainsorrel			1	1	
	Persicaria lapathifolia	curlytop knotweed	1				
	Polygonum amphibium	water knotweed	1	1			
	Polygonum aviculare subsp. *depressum*	prostrate knotweed	1	1	1		x
	Polygonum douglasii	Douglas' knotweed	1	1			
	Polygonum ramosissimum	bushy knotweed	1	1			
	Rumex acetosella	common sheep sorrel	1	1	1		x
	Rumex crispus	curly dock	1	1			x
	Rumex mexicanus	Mexican dock	1	1			
	Rumex densiflorus var. *pycnanthus*	dense-flowered dock	1				
POTAMOGETONACEAE	*Potamogeton natans*	floating pondweed		1			
	Potamogeton pusillus	small pondweed	1	1	1		
PRIMULACEAE	*Androsace septentrionalis*	pygmyflower rockjasmine	1	1	1	1	
	Primula parryi	Parry's primrose			1	1	
RANUNCULACEAE	*Aconitum columbianum*	Columbian monkshood	1	1	1		
	Anemone multifida	Pacific anemone		1			
	Aquilegia chrysantha	golden columbine	1	1	1	1	
	Aquilegia coerulea var. *pinetorum*	Colorado blue columbine	1	1			
	Delphinium geraniifolium	Clark Valley larkspur	1	1			
	Delphinium scaposum	tall mountain larkspur	1				
	Myosurus apetalus var. *montanus*	bristly mousetail		1			
	Ranunculus aquatilis var. *diffusus*	threadleaf crowfoot	1				
	Ranunculus cardiophyllus	heartleaf buttercup	1	1			
	Ranunculus eschscholtzii ●	Eschscholtz's buttercup			1		
	Ranunculus inamoenus var. *subaffinis* ●*	graceful buttercup	1	1	1	1	
	Ranunculus macounii	Macoun's buttercup	1	1			
	Ranunculus pedatifidus var. *affinis*	northern buttercup				1	
	Thalictrum fendleri	Fendler's meadow-rue	1	1	1		
RHAMNACEAE	*Ceanothus fendleri*	Fendler's buckbrush	1	1	1		
ROSACEAE	*Amelanchier alnifolia*	serviceberry	1	1			
	Amelanchier utahensis	Utah serviceberry	1	1			
	Chamaebatiaria millefolium	fernbush	1				
	Fallugia paradoxa	Apache plume	1				
	Fragaria vesca	woodland strawberry		1	1		
	Fragaria virginiana subsp. *glauca*	Virginia strawberry	1	1	1	1	
	Geum rossii ●	Ross' avens			1	1	
	Geum triflorum	old man's whiskers		1	1		
	Holodiscus dumosus	rockspirea	1	1	1		
	Potentilla anserina	silver cinquefoil	1	1			
	Potentilla concinna	elegant cinquefoil		1		1	
	Potentilla crinita	bearded cinquefoil	1	1	1		
	Potentilla diversifolia	varileaf cinquefoil			1	1	
	Potentilla fruticosa	shrubby cinquefoil	1				
	Potentilla hippiana	woolly cinquefoil	1	1	1	1	
	Potentilla jepsonii ●	Jepson's cinquefoil			1	1	
	Potentilla nivea ●	snow cinquefoil				1	
	Potentilla norvegica	Norwegian cinquefoil	1	1	1		
	Potentilla subviscosa	Navajo cinquefoil	1	1	1	1	
	Potentilla thurberi	scarlet cinquefoil	1	1			
	Prunus emarginata	bitter cherry		1			
	Purshia stansburiana	Stansbury cliffrose	1				
	Rosa woodsii var. *arizonica*	Woods rose	1	1			
	Rubus idaeus	American red raspberry		1	1		
	Sanguisorba minor	small burnet	1				x

Family	Scientific Name	Common Name	Ponderosa pine	Mixed conifer	Spruce-fir	Tundra	Nonnative
	Sibbaldia procumbens	creeping sibbaldia			1	1	
	Sorbus scopulina	Arizona mountain ash		1	1		
RUBIACEAE	*Galium wrightii*	Wright's bedstraw	1				
	Houstonia wrightii	pygmy bluet	1	1			
	Kelloggia galioides	milk kelloggia	1	1			
SALICACEAE	*Populus tremuloides*	quaking aspen	1	1	1		
	Salix bebbiana	Bebb willow	1	1	1		
	Salix lasiolepis	arroyo willow	1	1	1		
	Salix scouleriana	Scouler's willow		1	1		
SANTALACEAE	*Arceuthobium douglasii*	Douglas fir dwarf mistletoe		1	1		
	Arceuthobium microcarpum subsp. *aristatae* ● *	bristlecone pine dwarf mistletoe			1		
	Arceuthobium microcarpum subsp. *microcarpum*	western spruce dwarf mistletoe			1		
	Arceuthobium vaginatum subsp. *cryptopodum*	southwestern dwarf mistletoe	1				
SAPINDACEAE	*Acer glabrum*	Rocky Mountain maple		1			
SAXIFRAGACEAE	*Heuchera rubescens*	pink alumroot	1	1	1	1	
	Saxifraga caespitosa ●	tufted alpine saxifrage				1	
	Saxifraga flagellaris ●	whiplash saxifrage				1	
	Saxifraga rhomboidea	diamondleaf saxifrage	1	1	1	1	
	Saxifraga rivularis ●	weak saxifrage			1	1	
SCROPHULARIACEAE	*Limosella acaulis*	Owyhee mudwort		1			
	Limosella aquatica	water mudwort		1			
	Verbascum thapsus	common mullein	1	1	1		x
SOLANACEAE	*Nicotiana attenuata*	coyote tobacco	1				
	Physalis hederifolia	ivyleaf groundcherry	1				
	Physalis neomexicana	New Mexican groundcherry	1				
	Solanum triflorum	cutleaf nightshade	1				
TYPHACEAE	*Typha latifolia*	broadleaf cattail	1				
VERBENACEAE	*Verbena bracteata*	bigbract verbena	1				x
	Verbena macdougalii	MacDougal verbena	1	1			
VIOLACEAE	*Viola canadensis*	Canadian white violet	1	1	1	1	
	Viola nephrophylla	northern bog violet		1			
No. of families: 97	# Flowering = 508 # Liverworts = 2 # Mosses = 30 # Equisetum = 1 # Conifers = 12 # Ferns = 14 567 total		363	342	231	96	51

● only occurrence in Arizona is in the San Francisco Peaks = 33
* endemic to the San Francisco Peaks = 3

— APPENDIX II —

Mushrooms & Other Fungi of the San Francisco Peaks

Appendix II. Mushrooms and other Fungi of the Peaks. Courtesy of Marylou Fairweather and Edward Smith. P-J 5,000-6,000 feet, Ponderosa pine 6,000-8,500 feet, Mixed conifer 8,500-9,500 feet, Spruce-fir 9,500-11,500 feet, Tundra ~ > 11,500 feet. Aspens, of course occur in both Mixed conifer and Spruce-fir zones

Scientific Name	Common Name	Forest Type	Elevation
SAPROPHYTES (decomposers)			
Agaricus campestris	Field Mushroom	Meadow Mixed conifer	8,500-9,500 feet
Agaricus arvensis	Anise Mushroom	Mixed conifer	8,500-9,500 feet
Agaricus silvicola	Forest Agaric	Mixed conifer	8,500-9,500 feet
Auricularia auricula	Tree Ear	Mixed conifer Spruce-fir	8,500-9,500 feet 9,500-11,500 feet
Calvatia booniana	Giant Puffball	Ponderosa pine	6,000-8,500 feet
Chlorophyllum molybdites	Green Parasol	Urban and suburban lawns	<7,000 feet
Clavaria purpurea	Purple Coral	Mixed conifer Spruce-fir	8,500-9,500 feet 9,500-11,500 feet
Clavariadelphus borealis		Mixed conifer Spruce-fir	8,500-9,500 feet 9,500-11,500 feet
Clavulina cristata	Earth Coral	Mixed conifer	8,500-9,500 feet
Clavicorona pyxidata	Log Coral	Mixed conifer	8,500-9,500 feet
Coprinus atrementarius (don't eat w/ alcohol)	Inky Cap	Aspen Mixed conifer	 8,500-9,500 feet
Coprinus comatus	Shaggy Mane	Urban lawns, P-J, PIPO Mixed conifer Spruce-fir	 8,500-9,500 feet 9,500-11,500 feet
Coprinus micaceus	Inky Cap	Mixed conifer	8,500-9,500 feet
Coprinus nevius		Mixed conifer	8,500-9,500 feet
Cyathus striatus	Bird's nest	Mixed conifer Spruce-fir	8,500-9,500 feet 9,500-11,500 feet
Cystoderma fallax group		All conifer forest	
Dacrymyces palmatus	Pine Butter	Ponderosa pine	6,000-8,500 feet
Discina perlatum	Pig's Ear	Mixed conifer	8,500-9,500 feet
Fuligo septica	Slime Mold	All types	
Flammulina populicola	Velvet foot	Aspen Mixed conifer	 8,500-9,500 feet
Gloeophyllum sepiarium	(Brown Rot)	Mixed conifer	8,500-9,500 feet
Gymnopilus spectabilis	Laughing Mushroom	Ponderosa pine	6,000-8,500 feet

Scientific Name	Common Name	Forest Type	Elevation
Gyromitra gigas		Ponderosa pine & mixed conifer	6,000-9,500 feet
Gyromitra infula		Ponderosa pine & mixed conifer	6,000-9,500 feet
Helvella crispa		Mixed conifer	8,500-9,500 feet
Helvella lacunosa		Mixed conifer	8,500-9,500 feet
Hericium ramosum	Bear's Head	Aspen Mixed conifer	 8,500-9,500 feet
Hydnellum scrobiculatum		Mixed conifer	8,500-9,500 feet
Hydnum (Dentinum) repandum	Hedgehog	Aspen Mixed conifer	 8,500-9,500 feet
Lentinus lepideus		Ponderosa pine	6,000-8,500 feet
Lentinus ponderosus	Giant Sawtooth	Ponderosa pine	6,000-8,500 feet
Lycogala epidendrum	Log Slimer	Aspen Mixed conifer	 8,500-9,500 feet
Lycoperdon perlatum	Puffball	Ponderosa pine Mixed conifer Spruce-fir	6,000-8,500 feet 8,500-9,500 feet 9,500-11,500 feet
Morchella conica	Black Morel	Mixed conifer	8,500-9,500 feet
Mycena pura	Fragile Mycena	Mixed conifer	8,500-9,500 feet
Panaeolus campanulatus	Dung Bell	Mixed conifer	8,500-9,500 feet
Peziza repanda	Smoke Cup	Mixed conifer	8,500-9,500 feet
Phallus impudicus	Stinkhorn	Urban landscapes, also Aspen - Mixed conifer	 8,500-9,500 feet
Phyllotopsis nidulans		Aspen - Mixed conifer	8,500-9,500 feet
Pleurotus ostreatus	Oyster Mushroom	Aspen- Ponderosa pine Aspen - Mixed conifer	6,000-8,500 feet 8,500-9,500 feet
Pluteus admirabilis	(Salmon colored spores)	Aspen - Mixed conifer	8,500-9,500 feet
Pluteus cervinus		Gambel oak, Aspen, Mixed conifer	8,500-9,500 feet
Pycnoporellus alboluteus		Mixed conifer Spruce-fir	8,500-9,500 feet 9,500-11,500 feet
Psilocybe coprophila	Dung 'Shroom	Ponderosa pine	6,000-8,500 feet
Ramaria formosa	Earth Coral	Mixed conifer Spruce-fir	8,500-9,500 feet 9,500-11,500 feet
Sacrosphaera crassa		Ponderosa pine Mixed conifer	6,000-8,500 feet 8,500-9,500 feet
Spathularia flavida	Tongue Coral	Mixed conifer	8,500-9,500 feet
Stropharia kauffmanii		Mixed conifer	8,500-9,500 feet
Stropharia semiglobata	Dung Dome	Ponderosa pine Mixed conifer	6,000-8,500 feet 8,500-9,500 feet
Tremella mesenterica	Witch's Butter	Ponderosa pine Mixed conifer	6,000-8,500 feet 8,500-9,500 feet
Tricholomopsis platyphylla	Flattop woodrot	Mixed conifer	8,500-9,500 feet
Xeromphalina campanella	Conifer bell	Ponderosa pine	6,000-8,500 feet
PARASITES (pathogens)			
Armillaria ostoyae (Both parasite and saprophyte)	Honey Mushroom	Ponderosa pine Mixed conifer Spruce-fir	6,000-8,500 feet 8,500-9,500 feet 9,500-11,500 feet
Cronartium arizonicum	Limb Rust	Ponderosa pine Mixed conifer	6,000-8,500 feet 8,500-9,500 feet
Cryptoporus volvatus	Pouch Fungus	P-J	5,000-6,000 feet

Scientific Name	Common Name	Forest Type	Elevation
		Ponderosa pine	6,000-8,500 feet
		Mixed conifer	8,500-9,500 feet
		Spruce-fir	9,500-11,500 feet
Fomitopsis pinicola	Red Belt	Ponderosa pine	6,000-8,500 feet
(Both parasite and saprophyte)		Mixed conifer	8,500-9,500 feet
		Spruce-fir	9,500-11,500 feet
Ganoderma applanatum	Artist's Conk	Aspen-Ponderosa pine	6,000-8,500 feet
		Aspen-Mixed conifer	8,500-9,500 feet
		Aspen-Spruce-fir	9,500-11,500 feet
Hypomyces luteovirens on *Russula emetica*		Ponderosa pine	6,000-8,500 feet
		Mixed conifer	8,500-9,500 feet
		Spruce-fir	9,500-11,500 feet
Hypomyces lactifluorom on *Russula brevipes*	Lobster	Ponderosa pine	6,000-8,500 feet
		Mixed conifer	8,500-9,500 feet
Onnia tomentosa	Wooly velvet polyphore	Mixed conifer	8,500-9,500 feet
		Spruce-fir	9,500-11,500 feet
Melampsora sp.	Leaf Rust	Aspen-Ponderosa pine	6,000-8,500 feet
		Aspen-Mixed conifer	8,500-9,500 feet
Phellinus schweinitzii	Cow Patty Fungus	Ponderosa pine	6,000-8,500 feet
		Mixed conifer	8,500-9,500 feet
		Spruce-fir	9,500-11,500 feet
Phellinus tremulae	False Tinder Conk	Aspen-Ponderosa pine	6,000-8,500 feet
		Aspen-Mixed conifer	8,500-9,500 feet
		Aspen-Spruce-fir	9,500-11,500 feet
Pholiota squarrosa	Basque Garlic	Aspen-Ponderosa pine	6,000-8,500 feet
		Aspen-Mixed conifer	8,500-9,500 feet
		Aspen-Spruce-fir	9,500-11,500 feet
Sparassis crispa	Cauliflower	Ponderosa pine	6,000-8,500 feet
		Mixed conifer	8,500-9,500 feet
		Spruce-fir	9,500-11,500 feet
Inonotus tomentosus		Mixed conifer	8,500-9,500 feet
		Spruce-fir	9,500-11,500 feet
MYCORRHIZAE (mutualist symbionts)			
Amanita caesarea	Caeser's Amanita	Ponderosa pine	6,000-8,500 feet
Amanita gemmata		Mixed conifer	8,500-9,500 feet
Amanita muscaria	Fly Agaric	Ponderosa pine	6,000-8,500 feet
		Mixed conifer	8,500-9,500 feet
		Spruce-fir	9,500-11,500 feet
Amanita pantherina		Mixed conifer	8,500-9,500 feet
Amanita virosa	Destroying Angel	Mixed conifer	8,500-9,500 feet
Amanita vaginata	Grisette	Mixed conifer	8,500-9,500 feet
Armillaria straminea		Mixed conifer	8,500-9,500 feet
Boletus barrowsii	White King Bolete	Ponderosa pine	6,000-8,500 feet
		Mixed conifer	8,500-9,500 feet
Boletus edulis	Red King Bolete	Ponderosa pine	6,000-8,500 feet
		Mixed conifer	8,500-9,500 feet
		Spruce-fir	9,500-11,500 feet
Cantharellus spp.	Chanterelle	Ponderosa pine	6,000-8,500 feet
		Mixed conifer	8,500-9,500 feet
		Spruce-fir	9,500-11,500 feet
Clitocybe gibba		Mixed conifer	8,500-9,500 feet

Scientific Name	Common Name	Forest Type	Elevation
Cortinarius Glaucopus (group)		Mixed conifer Spruce-fir	8,500-9,500 feet 9,500-11,500 feet
Cortinarius spp.	Snowbank fungus	Mixed conifer Spruce-fir	8,500-9,500 feet 9,500-11,500 feet
Hebeloma insigne		Mixed conifer	8,500-9,500 feet
Hydnum imbricatum	Hawk Wing	Mixed conifer	8,500-9,500 feet
Hygrophoropsis aurantiaca	False Chanterelle	Mixed conifer	8,500-9,500 feet
Hygrophorus chrysodon	Golden Waxer	Mixed conifer	8,500-9,500 feet
Hygrophorus conicus	Red Cone	Mixed conifer	8,500-9,500 feet
Hygrophorus erubescens	Pine waxer	Mixed conifer	8,500-9,500 feet
Hygrophorus speciosus	Orange Slimer	Mixed conifer	8,500-9,500 feet
Lactarius deliciosus	Milk Cap	Mixed conifer	8,500-9,500 feet
Leccinum aurantiacum	Aspen Bolete	Mixed conifer	8,500-9,500 feet
Leucopaxillus amarus	Bitter Fairy Ring	Mixed conifer	8,500-9,500 feet
Leucopaxillus albissimus	White Fairy Ring	Mixed conifer	8,500-9,500 feet
Russula claroflava	Yellow-Ochre	Mixed conifer	8,500-9,500 feet
Russula emetica	Hurl Cap	Mixed conifer	8,500-9,500 feet
Russula brevipes	Short Foot	Mixed conifer	8,500-9,500 feet
Russula rosacea	Rosy Stipe	Mixed conifer	8,500-9,500 feet
Russula variata		Mixed conifer	8,500-9,500 feet
Suillus brevipes		Mixed conifer	8,500-9,500 feet
Suillus kaibabensis (on ponderosa pine)	Slippery Jack	Ponderosa pine Mixed conifer	6,000-8,500 feet 8,500-9,500 feet
Suillus lakei (on douglas-fir)		Mixed conifer	8,500-9,500 feet
Suillus sibiricus (on 5-needle white pine)		Mixed conifer	8,500-9,500 feet
Suillus tomentosus		Mixed conifer	8,500-9,500 feet

— APPENDIX III —

Birds of the San Francisco Peaks

Appendix III. Birds of the Peaks. With assistance from John Coons, S.O. Williams, Steven Carothers, Chuck LaRue, C.H. Merriam (1890) and Ebird.org.

Famiy	Scientific name	Common name
Anatidae	*Spatula discors*	Blue-winged Teal
	Spatula cyanoptera	Cinnamon Teal
	Anas platyrhynchos	Mallard
Phasianidae	*Dendragapus obscurus*	Dusky Grouse
	Meleagris gallopavo	Wild Turkey
Columbidae	*Patagioenas fasciata*	Band-tailed Pigeon
	Streptopelia decaocto	Eurasian Collared-Dove
	Zenaida macroura	Mourning Dove
Cuculidae	*Geococcyx californianus*	Greater Roadrunner
Caprimulgidae	*Chordeiles minor*	Common Nighthawk
	Phalaenoptilus nuttallii	Common Poorwill
Apodidae	*Aeronautes saxatalis*	White-throated Swift
Trocilidae	*Archilochus alexandri*	Black-chinned Hummingbird
	Calypte anna	Anna's Hummingbird
	Selasphorus platycercus	Broad-tailed Hummingbird
	Selasphorus rufus	Rufous Hummingbird
	Selasphorus calliope	Calliope Hummingbird
Rallidae	*Fulica americana*	American Coot
Recurvirostridae	*Recurvirostra americana*	American Avocet
Charadriidae	*Aegialitis vocifera*	Killdeer
Scolopacidae	*Calidris bairdii*	Baird's Sandpiper
	Calidris minutilla	Least Sandpiper
	Actitis macularia	Spotted Sandpiper
	Tringa solitaria	Solitary Sandpiper
	Pharalopus lobatus	Red-necked Phalarope
Cathartidae	*Cathartes aura*	Turkey Vulture
Accipitridae	*Haliaeetus leucocephalus*	Bald Eagle
	Circus hudsonius	Northern Harrier

	Accipiter velox	Sharp-shinned Hawk
	Accipiter cooperi	Cooper's Hawk
	Accipiter atricapillus	Northern Goshawk
	Buteo swainsoni	Swainson's Hawk
	Buteo albonotatus	Zone-tailed Hawk
	Buteo jamaicensis	Red-tailed Hawk
	Buteo regalis	Ferruginous Hawk
	Aquila chrysaetos	Golden Eagle
Strigidae	*Bubo virginianus*	Great Horned Owl
	Glaucidium gnoma	Northern Pygmy-Owl
	Strix occidentalis	Spotted Owl
	Asio otus	Long-eared Owl
	Aegolius acadicus	Northern Saw-whet Owl
Picadae	*Melanerpes lewis*	Lewis's Woodpecker
	Melanerpes formicivorus	Acorn Woodpecker
	Sphyrapicus thyroideus	Williamson's Sapsucker
	Sphyrapicus nuchalis	Red-naped Sapsucker
	Picoides pubescens	Downy Woodpecker
	Picoides villosus	Hairy Woodpecker
	Picoides dorsalis	American Three-toed Woodpecker
	Colaptes auratus	Northern Flicker
Falconidae	*Falco sparverius*	American Kestrel
	Falco columbarius	Merlin
	Falco peregrinus	Peregrine Falcon
	Falco mexicanus	Prairie Falcon
Tyrannidae	*Contopus cooperi*	Olive-sided Flycatcher
	Contopus sordidulus	Western Wood-pewee
	Empidonax hammondii	Hammond's Flycatcher
	Empidonax wrightii	Gray Flycatcher
	Empidonax oberholseri	Dusky Flycatcher
	Empidonax occidentalis	Cordilleran Flycatcher
	Sayornis nigricans	Black Phoebe
	Sayornis saya	Say's Phoebe
	Myiarchus cinerascens	Ash-throated Flycatcher
	Tyrannus vociferans	Cassin's Kingbird
Laniidae	*Lanius ludovicianus*	Loggerhead Shrike
Vireonidae	*Vireo vicinior*	Gray Vireo
	Vireo cassinii	Cassin's Vireo
	Vireo plumbeus	Plumbeous Vireo
	Vireo gilvus	Warbling Vireo
Corvidae	*Gymnorhinus cyanocephalus*	Pinyon Jay
	Cyanocitta stelleri	Steller's Jay
	Aphelocoma woodhouseii	Woodhouse's Scrub-Jay
	Nucifraga columbiana	Clark's Nutcracker
	Corvus brachyrhynchos	American Crow
	Corvus corax	Common Raven
Hirundinidae	*Tachycineta bicolor*	Tree Swallow
	Tachycineta thalassina	Violet-green Swallow
	Stelgidopteryx serripennis	Northern Rough-winged Swallow
	Petrochelidon pyrrhonota	Cliff Swallow
	Hirundo rustica	Barn Swallow
Paridae	*Poecile gambeli*	Mountain Chickadee
	Baeolophus ridgwayi	Juniper Titmouse

Aegithalidae	*Psaltriparus minimus*	Bushtit
Sittadae	*Sitta canadensis*	Red-breasted Nuthatch
	Sitta carolinensis	White-breasted Nuthatch
	Sitta pygmaea	Pygmy Nuthatch
Certhiidae	*Certhia americana*	Brown Creeper
Troglodytidae	*Salpinctes obsoletus*	Rock Wren
	Catherpes mexicanus	Canyon Wren
	Troglodytes aedon	House Wren
	Thryomanes bewickii	Bewick's Wren
Polioptilidae	*Polioptila caerulea*	Blue-gray Gnatcatcher
Cinclidae	*Cinclus mexicanus*	American Dipper
Regulidae	*Regugus satrapa*	Golden-crowned Kinglet
	Regulus calendula	Ruby-crowned Kinglet
Turdidae	*Sialia mexicana*	Western Bluebird
	Sialia currucoides	Mountain Bluebird
	Myadestes townsendi	Townsend's Solitaire
	Catharus ustulatus	Swainson's Thrush
	Catharus guttatus	Hermit Thrush
	Turdus migratorius	American Robin
Mimidae	*Dumetella carolinensis*	Gray Catbird
	Oreoscoptes montanus	Sage Thrasher
	Mimus polyglottos	Northern Mockingbird
Sturnidae	*Sturnus vulgaris*	Eurasian Starling
Bombycillidae	*Bombycilla cedrorum*	Cedar Waxwing
Peucedramidae	*Peucedramus taeniatus*	Olive Warbler
Passeridae	*Passer domesticus*	House Sparrow
Motacillidae	*Anthus rubescens*	American Pipit
Fringillidae	*Coccothraustes vespertinus*	Evening Grosbeak
	Pinicola enucleator	Pine Grosbeak
	Leucosticte tephrocotis	Gray-crowned Rosy-Finch
	Leucosticte atrata	Black Rosy-Finch
	Carpodacus mexicanu	House Finch
	Carpodacus cassinii	Cassin's Finch
	Loxia curvirostra	Red Crossbill
	Carduelis pinus	Pine Siskin
	Carduelis psaltria	Lesser Goldfinch
	Carduelis lawrencei	Lawrence's Goldfinch
	Carduelis tristis	American Goldfinch
Passerellidae	*Pipilo chlorurus*	Green-tailed Towhee
	Pipilo maculatus	Spotted Towhee
	Spizella passerina	Chipping Sparrow
	Spizella breweri	Brewer's Sparrow
	Pooectes gramineus	Vesper Sparrow
	Chondestes grammacus	Lark Sparrow
	Amphispiza nevadensis	Sagebrush Sparrow
	Passerculus sandwichensis	Savannah Sparrow
	Passerlla iliaca	Fox Sparrow

	Melospiza melodia	Song Sparrow
	Melospiza lincolnii	Lincoln's Sparrow
	Zonotrichia leucophrys	White-crowned Sparrow
	Junco hyemalis	Dark-eyed Junco
Icteridae	*Sturnella neglecta*	Western Meadowlark
	Icterus bullockii	Bullock's Oriole
	Icterus parisorum	Scott's Oriole
	Agelaius phoeniceus	Red-winged Blackbird
	Molothrus ater	Brown-headed Cowbird
	Euphagus cyanocephalus	Brewer's Blackbird
	Quiscalus mexicanus	Great-tailed Grackle
Parulidae	*Seiurus aurocapillus*	Ovenbird
	Vermivora celata	Orange-crowned Warbler
	Vermivora luciae	Lucy's Warbler
	Vemivora ruficapilla	Nashville Warbler
	Vermivora virginiae	Virginia's Warbler
	Oporornis tolmiei	MacGillivray's Warbler
	Oporrornis formosus	Kentucky Warbler
	Geothlypis trichas	Common Yellowthroat
	Dendroica petechia	Yellow Warbler
	Dendroica pensylvanica	Chestnut-sided Warbler
	Dendroica coronata	Yellow-rumped Warbler
	Dendroica graciae	Grace's Warbler
	Dendroica nigrescens	Black-throated Gray Warbler
	Dendroica townsendi	Townsend's Warbler
	Dendroica occidentalis	Hermit Warbler
	Wilsonia pusilla	Wilson's Warbler
	Cardellina rubrifrons	Red-faced Warbler
Cardinalidae	*Piranga flava*	Hepatic Tanager
	Piranga ludoviciana	Western Tanager
	Pheucticus melanocephalus	Black-headed Grosbeak
	Passerina amoena	Lazuli Bunting
	Passerina cyanea	Indigio Bunting

Total = 161 species

— APPENDIX IV —

Mammals of the San Francisco Peaks

Appendix IV. Mammals of the Peaks. With assistance from Tad Theimer, NAU, Carol Chambers, NAU, D. F. Hoffmeister (1986) and C. H. Merriam (1890).

Famiy	Scientific name	Common name
Antilocapridae	*Antilocapra americana*	pronghorn
Canidae	*Canis latrans*	coyote
	Urocyon cinereoargenteus	gray fox
Cervidae	*Cervus elephas*	elk
	Odocoileus hemionus	mule deer
	Odocoileus virginianus	white-tailed deer
Erethizontidae	*Erethizon dosrsatum*	yellow-haired porcupine
Felidae	*Felis concolor*	mountain lion
	Lynx baileyi	bobcat
Geomyidae	*Thomomys bottae*	Botta's pocket gopher
Heteromyidae	*Dipodomys ordii*	Ord's kangaroo rat
	Perognathus flavus	silky pocket mouse
Leporidae	*Lepus californicus*	black-tailed jackrabbit
	Sylviilagus nuttallii	Nuttall's cottontail
Mephitidae	*Mephitis mephitis*	striped skunk
	Spilogale gracilis	western spotted skunk
Muridae	*Microtus mogollonensis*	Mogollon vole
	Neotoma mexicana	Mexican wood rat
	Onychomys leucogaster	northern grasshopper mouse
	Peromyscus maniculatus	deer mouse
	Peromyscus trueii	pinyon mouse
Mustelidae	*Taxidea taxus*	badger
Procyonidae	*Bassariscus astutus*	ringtail
Sciuridae	*Callospermophilus lateralis*	golden-mantled ground squirrel
	Cynomys gunnisoni	prairie dog
	Otospermophilus variegatus	rock squirrel
	Sciurus aberti	Abert's squirrel
	Tamias cinereicollis	grey-collared chipmunk

	Tamias dosalis	cliff chipmunk
	Tamiasciurus hudsonicus	red squirrel
	Xerospermophilus spilosoma	spotted ground squirrel
Soricidae	*Sorex monticolus*	mountain shrew
	Sorex nanus	dwarf shrew
	Sorex merriami	Merriam's shrew
Ursidae	*Ursus americanus*	black bear
Vespertilionidae	*Eptesicus fuscus*	big brown bat
	Myotis volans	Long-legged myotis
	Myotis occultus	Arizona myotis
	Myotis evotis	Long-eared myotis
	Myotis auriculus	Southwestern myotis

— APPENDIX V —

Plants on Mt. Elden and locations other than the San Francisco Peaks

Appendix V. Plant list for nearby Mt. Elden. This list was developed by Max Licher and Gwen Waring, working from lists provided by William Moir, the Deaver Herbarium, NAU, and SEINET.

Family	Scientific Name	Common Name	Ponderosa pine	Mixed conifer	Spruce-fir	Nonnative
SPIKEMOSSES						
SELAGINELLACEAE	*Selaginella underwoodii*	Underwood's spikemoss	1			
HORSETAILS						
EQUISETACEAE	*Equisetum arvense*	field horsetail		1		
FERNS						
ASPLENIACEAE	*Asplenium adiantum-nigrum***	black spleenwort	1			
	*Asplenium platyneuron***	ebony spleenwort	1			
	Asplenium resiliens	blackstem spleenwort	1			
DRYOPTERIDACEAE	*Polystichum scopulinum*	mountain hollyfern	1	1		
PTERIDACEAE	*Cheilanthes eatonii*	Eaton's lipfern	1			
	Cheilanthes feei	slender lipfern	1			
	Cheilanthes fendleri	Fendler's lipfern	1	1		
	Cheilanthes wootoni	beaded lipfern	1			
	Pellaea atropurpurea	purple cliffbrake	1			
	Pellaea wrightiana	Wright's cliffbrake	1			
	Pentagramma triangularis	goldback fern	1	1		
WOODSIACEAE	*Cystopteris utahensis*	Utah bladderfern	1			
CONIFERS						
CUPRESSACEAE	*Juniperus deppeana*	alligator juniper	1			
	Juniperus monosperma	oneseed juniper	1			
	Juniperus scopulorum	Rocky Mountain juniper	1	1		
FLOWERING PLANTS						
AMARANTHACEAE	*Amaranthus albus*	prostrate pigweed		1		x
	Atriplex canescens	fourwing saltbush	1			
	Atriplex rosea	tumbling saltweed	1			
	Chenopodium rubrum	chenopodium		1		
	Chenopodium incanum	mealy goosefoot	1	1		
	Krascheninnikovia lanata	winterfat	1			
ANACARDIACEAE	*Rhus glabra*	smooth sumac	1			
	Rhus trilobata	skunkbush sumac	1			
	Toxicodendron rydbergii	western poison ivy	1			
APOCYNACEAE	*Asclepias tuberosa*	butterfly milkweed	1			

			Ponderosa pine	Mixed conifer	Spruce-fir	Nonnative
ASPARAGACEAE	*Agave parryi*	Parry's agave	1			
	Yucca baccata	Bannana yucca	1			
ASTERACEAE	*Ambrosia acanthicarpa*	ragweed	1			
	Ambrosia psilostachya	Cuman ragweed	1			
	Artemisia biennis	biennial wormwood	1	1		x
	Artemisia campestris	field sagewort	1			
	Baccharis pteronioides	yerba de pasmo	1			
	Chrysothamnus depressus	longflower rabbitbrush	1			
	Erigeron canus	hoary fleabane	1			
	Erigeron modestus	plains fleabane	1			
	Erigeron saxatilis	rock fleabane		1		
	Gaillardia pinnatifida	red dome blanketflower	1			
	Hymenopappus filifolius var. *lugens*	Idaho hymenopappus	1			
	Hymenoxys acaulis	Arizona Actinea	1			
	Lactuca tatarica	blue lettuce	1	1		
	Laennecia sophiifolia	leafy marshtail	1			
	Perityle ciliata	fringed rockdaisy	1			
	Pseudognaphalium canescens	Wright's cudweed		1	1	
	Senecio flaccidus	threadleaf ragwort	1			
	Symphyotrichum lanceolatum	white panicle aster	1	1		
	Townsendia exscapa	stemless Townsend daisy	1			
	Tragopogon porrifolius	salsify	1			x
	Uropappus lindleyi	silver puffs	1			
BERBERIDACEAE	*Berberis fremontii*	Fremont's barberry	1			
	Berberis haematocarpa	bloodberry barberry	1			
BORAGINACEAE	*Cryptantha fendleri*	sanddune cryptantha	1			
BRASSICACEAE	*Boechera gracilipes*	Flagstaff rockcress				
	Boechera perennans	perennial rockcress	1			
	Chorispora tenella	blue mustard	1			x
	Descurainia sophia	herb sophia	1			x
	Erysimum capitatum	sanddune wallflower	1			
	Erysimum repandum	spreading wallflower	1			
	Sisymbrium altissimum	tall tumblemustard	1			x
	Thelypodium wrightii	Wright's thelypody	1	1		
CACTACEAE	*Echinocereus coccineus*	scarlet hedgehog cactus	1			
	Opuntia engelmannii var. *engelmannii*	cactus apple	1			
	Opuntia macrorhiza	twistspine pricklypear	1			
COMMELINACEAE	*Tradescantia pinetorum*	pinewoods spiderwort	1			
CUCURBITACEAE	*Cucurbita foetidissima*	Missouri gourd	1			
ERICACEAE	*Pyrola picta*	whiteveined wintergreen		1	1	
FABACEAE	*Astragalus tephrodes*	ashen milkvetch	1			
	Psoralidium lanceolatum	lemon scurfpea	1			
	Trifolium gymnocarpon	hollyleaf clover	1			
	Trifolium repens	white clover	1	1		
FAGACEAE	*Quercus undulata*	Wavyleaf Oak	1			
GARRYACEAE	*Garrya wrightii*	Wright's silktassel	1			
GROSSULARIACEAE	*Ribes inerme*	whitestem gooseberry	1			
JUGLANDIACEAE	*Juglans major*	Arzona walnut	1			
LAMIACEAE	*Hedeoma oblongifolia*	oblongleaf false pennyroyal	1			
	Monarda pectinata	pony beebalm	1			
	Salvia subincisa	sawtooth sage	1			
	Stachys rothrockii	Rothrock's hedgenettle	1			
LINACEAE	*Linum neomexicanum*	New Mexico yellow flax	1			
LOASACEAE	*Mentzelia albicaulis*	whitestem blazingstar	1			
	Mentzelia montana	Montana blazingstar	1			
MALVACEAE	*Malva neglecta*	common mallow	1	1		x
NYCTAGINACEAE	*Mirabilis oxybaphoides*	smooth spreading four o'clock	1			
OLEACEAE	*Fraxinus velutina*	velvet ash	1			

			Ponderosa pine	Mixed conifer	Spruce-fir	Nonnative
ONAGRACEAE	*Gaura hexandra* subsp. *gracilis*	harlequinbush	1			
OROBANCHACEAE	*Orthocarpus purpureoalbus*	purplewhite owl's-clover	1			
	Pedicularis centranthera	dwarf lousewort	1			
PHRYMACEAE	*Mimulus rubellus*	little redstem monkeyflower	1			
PLANTAGINACEAE	*Callitriche heterophylla*	twoheaded water-starwort	1	1		
	Penstemon linarioides subsp. *compactifolius*	toadflax beardtongue	1			
	Penstemon nudiflorus	Flagstaff penestemon	1			
	Penstemon pseudospectabilis	desert penstemon	1			
	Penstemon rostriflorus	Bridge's penstemon	1			
POACEAE	*Achnatherum hymenoides*	Indian ricegrass	1			
	Aristida divaricata	poverty threeawn	1			
	Aristida purpurea var. *longiseta*	Fendler threeawn	1			
	Bromus japonicus	Japanese brome	1	1		
	Dichanthelium oligosanthes subsp. *scribnerianum*	Scribner's Panicgrass	1			
	Eragrostis curvula	weeping lovegrass	1			x
	Eragrostis pectinacea	tufted lovegrass	1			
	Lycurus setosus	common wolfstail	1			
	Monroa squarrosa	false buffalograss	1			
	Muhlenbergia longiligula	longtongue muhly	1			
	Muhlenbergia pauciflora	New Mexico muhly	1	1		
	Muhlenbergia tenuifolia (monticola)	mesa muhly	1			
	Panicum bulbosum	bulb panicgrass	1			
	Piptatherum micranthum	littleseed ricegrass	1			
	Sporobolus cryptandrus	sand dropseed	1			
	Sporobolus interruptus	black dropseed	1			
POLEMONIACEAE	*Gilia opthalmoides*	eyed gilia	1			
	Phlox austromontana	mountain phlox	1	1		
	Phlox gracilis	slender phlox	1	1		
POLYGONACEAE	*Eriogonum alatum*	winged buckwheat	1			
	Eriogonum arcuatum	James' buckwheat	1			
	Eriogonum fasciculatum	Eastern Mojave buckwheat	1			
	Eriogonum wrightii	bastardsage	1			
	Polygonum convolvulus	black bindweed	1			
RANUNCULACEAE	*Aquilegia desertorum*	desert columbine	1	1		
ROSACEAE	*Cercocarpus montanus* var. *paucidentatus*	hairy mountain mahogany	1			
	Purshia stansburiana	Cliffrose	1			
RUBIACEAE	*Galium aparine*	stickywilly	1			
SOLANACEAE	*Datura wrightii*	sacred thorn-apple	1			
	Solanum jamesii	wild potato	1			
VITACEAE	*Vitis arizonica*	canyon grape	1	1	1	
	Total = 125 species		118	24	3	8

** = Mt. Elden only in Arizona

— GLOSSARY —

Alluvial, alluvium Sediment and organic debris left behind by water, such as stream deposits along a river.

Angiosperms Flowering vascular plants are distinguished from gymnosperms like conifers by producing seeds that are fully enclosed in the female ovary.

Aquifer An aquifer is a body of permeable rock that can contain or transmit water. The mighty Coconino Aquifer stores water deep under ground, while the Inner Basin Aquifer is really a glacial moraine comprised of unconsolidated mounds of rock and sediment perched on the floor of the Inner Basin.

Avalanche A mass of snow, ice and/or rock falling rapidly down a mountainside.

Boreal forest The coniferous forest belt of the Northern Hemisphere covers about 4.6 million miles^2; it borders the tundra in the north and mixed forests and grasslands in the south.

Browsing Feeding on the growing tips of low tree branches and shrubs, especially by mammals.

Carbohydrate The simplest form of carbohydrate produced by plant photosynthesis is glucose sugar, made from carbon dioxide and water.

Carbon 14 dating Radiocarbon dating was developed in the 1940's and Willard F. Libby won a Nobel Prize in Chemistry in 1960 for spearheading its discovery. This method has enabled scientists to date organic material back to almost 50,000 years. Carbon has three isotopes or forms: ^{12}C, ^{13}C, ^{14}C, with ^{14}C being the most unstable or radioactive. Plants and animals absorb all three carbon forms until they die; after that point, the proportion of unstable ^{14}C can be measured to estimate the age of organic remains such as pieces of plants found in packrat middens and cores of lake sediment in the western United States. Today, the half-life of ^{14}C is measured as 5,730±40 years.

Climax Species or communities representing either a prolonged stage or the final stage of ecological succession.

Cline A gradation of one or more traits along a gradient, such as elevation or latitude. Most western conifers occur as clines along these gradients, expressing variation in growth rates and cold tolerance.

Clone An organism that is produced asexually by its parent and is genetically identical to it. Aspen clones spread via underground stems.

Cold desert A dry corridor from Arizona to British Columbia. It typically receives less than 10 inches of precipitation per year, and, unlike the southern deserts, experiences freezing temperatures in winter.

Distribution Where a species lives.

DNA Deoxyribonucleic acid, the molecular

basis of heredity in all organisms except certain viruses. DNA encodes the amino acid sequence of enzymes and structural proteins in its sequence of four bases (thymine, adenine, cytosine and guanine).

Endemic, -ism A species whose range is restricted to a specified region or locality.

Evaporation The conversion of water molecules from a liquid to a gaseous form.

Evolution In a broad sense, the origin of entities possessing different genetic states of one or more characteristics, and changes in their proportions over time.

Extinct, -ion The total loss of a species.

Fault A fracture in the Earth's crust, along which movement or displacement has taken place, resulting from the action of plate tectonic forces.

Forbs Herbaceous or nonwoody plants with flowers.

Fluvial Fluvial processes are associated with rivers and streams, and the deposits and landforms created by them.

Gene The functional unit of heredity.

Grazing Animals, especially mammals, that feed on grasses and low flowers.

Herbivore An animal that feeds on plants.

Hybrid, hybridize An individual formed by mating between genetically different populations or species.

Invertebrates Animals such as insects that lack a backbone.

Island arc A curved line of volcanic islands, often located at a tectonic plate margin. Island arcs helped to build western North America.

Keystone species A species whose ecological presence is key to the success of other species within its community.

Marine Isotope Stages These refer to 2 oxygen isotopes whose ratios in oceans reveal climate patterns dating back several million years. The ratio of the heavier ^{18}O isotope increased in glacial periods, while that of the lighter ^{16}O isotope increased in warmer oceans.

Mesic More moderate environmental conditions, and particularly greater water availability.

Migration During the Pleistocene and Holocene, plants and animals migrated or shifted their ranges in response to cold and then warmth.

Natural selection The process whereby the frequency of heritable traits that better adapt a population to the environment increases from generation to generation, as a result of increased reproductive success of the individuals possessing those traits.

Orogeny Mountain building episode.

Packrat middens Packrats (*Neotoma spp.*) build nests of pieces of surrounding vegetation in caves and rock shelters from the Arctic to Nicaragua. They urinate on their nests, which crystallizes the contents and protects them from weathering. Contemporary nests may contain 75% or more of surrounding plant species, reflecting the nature of the surrounding community. These nests are abundant in the West and provide a fantastic record of western plant communities and their changes over the last nearly 50,000 years, using Carbon 14 dating techniques (see Carbon 14 dating).

Photosynthesis, photosynthesize Plants produce carbohydrates—the staff of life—in their leaves by absorbing CO_2 through leaf pores, energy from the sun, and water from the ground. The metabolic systems of most animals require the breakdown of carbohydrates for energy and growth. A convenient byproduct of this is atmospheric oxygen, on which most plants and animals also depend.

Plate tectonics A verified theory that the Earth's outer shell consists of a number of thick plates that are slowly moving and changing in size. Intense geologic activity, including faulting, earthquakes and volcanism, occurs at plate boundaries.

Pleistocene The Pleistocene Epoch occurred between about 2.6 Ma and 11,700 years ago. It was a cold period in the Earth's history, with 15 to 20 phases when glaciers expanded throughout the world. There are many records of plant and animal migrations during this time.

Pollination The transfer of pollen—the plant equivalent of sperm—to a receptive surface, usually the female stigma in plants, leading to pollination of eggs and the production of seeds.

Productivity The rate of production of new biomass, as in plant growth. Many plants, such as aspens and willows, are highly productive with fast growth rates.

Rain shadow Most atmospheric moisture is captured as its rises up over mountains, leaving little moisture for the lee side of these barriers. As air descends on the lee side of a mountain it is warmer and drier.

Range A species' distribution, where it lives.

Recruitment The increase in a natural population, as progeny establish and immigrants arrive.

Sister species Species derived from a common ancestor shared by no other species.

Species richness The number of species in a particular area.

Succession Progression of plant communities towards a climax or state in which the species are able to withstand the extremes of the climate and biotic interactions, and are relatively stable in their association, and in which the mature members of the community tend to be replaced by their own offspring.

Symbiotic, symbiosis The interaction between several species that is often, though not always, beneficial to both.

Taxa, taxon Biological groups, ranging from subspecies, varieties, etc., to genus or even family.

Tectonic forces Forces generated within the Earth that result in uplift, movement or deformation of part of the earth's crust.

Treeline Treeline is the edge of the habitat at which trees are able to grow. It is found at high elevations and high latitudes.

Vertebrate An animal whose spinal nerve cord is located inside a backbone, a column of bony segments called vertebrae.

Western Cordillera Mountain ranges and basins through western North America.

Wind pollination, dispersal Many plants, such as most conifers, rely on the wind to disperse their pollen from male to female cones, and then to disperse their seeds. By contrast, many modern flowering plants rely on animals to pollinate them and sometimes disperse their seeds.

Xeric Dry conditions.

— REFERENCES —

Adams, H.D & T.E. Kolb. 2005. Tree growth response to drought and temperature in a mountain landscape in northern Arizona, USA. Journal of Biogeography 32:1629-1640.

Adams, R.P., C.J. Earle & D. Thornburg. 2011. Taxonomy of infraspecific taxa in *Abies lasiocarpa*: leaf essential oils and DNA. Phytologia 93:87-101.

Aguirre-Planter, E., J.P Jaramillo-Correa, S. Gomez-Acevedo, D.K. Khasa, J. Buosquet & L.E. Eguiarte. 2011. Phylogeny, diversification rates and species boundaries of Mesoamerican firs (*Abies*, Pinaceae) in a genus-wide context. Molecular Genetics and Evolution 62:263-274.

Allen, C.D. 2002. Lots of lightning and plenty of people: an ecological history of fire in the upland Southwest. *In* Vale, T.R., Fire, native peoples, and the natural landscape. Island Press, Washington, DC.

Anderson, R.S. 1993. A 35,000 year vegetation and climate history from Potato Lake, Mogollon Rim, Arizona. Quaternary Research 40:351-359.

Anderson, R. S., J.I. Betancourt, J.I. Mead, R.H. Hevly & D.P. Adam. 2000. Middle- and late-Wisconsin paleobotanic and paleoclimatic records from the southern Colorado Plateau, USA. Paleogeography, Paleoclimatography, Paleoecology 155:31-57.

Ashworth, D. 1997. Biography of a small mountain. Small Mountain Books. Flagstaff, AZ.

Arnold, J.F. 1950. Changes in ponderosa pine bunchgrass ranges in northern Arizona resulting from pine regeneration and grazing. Journal of Forestry February:118-126.

Avery, C.A., L.R. Dexter, R.R. Wier, W.G. Delinger, A. Techle & R.J. Becker. 1993. Estimating and verifying the potential for evapo-sublimation losses from winter snow cover in Northern Arizona. P.O. 40- 82FT-2-0313, Rocky Mountain Forest and Range Experiment Station, U.S. Forest Service & Northern Arizona University, Flagstaff AZ.

Axelrod, D. I. 1990. Age and origins of the Cordillera flora. Journal of Biogeography 12:21-47.

Babbitt, B. 1964. A paleomagnetic chronology of the San Francisco volcanic field. Plateau 40:11-14.

Bailey, J.K & T.G. Whitham. 2002. Interactions among fire, aspens and elk affect insect diversity: a reversal of a community response. Ecology 82:1701-1712.

Barrows, J.S. 1978. Lightning fires in southwestern forests. Final Report. USDA USFS, Northern Forest Fire Laboratory, Intermountain Forest and Range Experiment Station, Cooperative Agreement 16-568, with Rocky Mountain Forest and Range Experiment Station, Fort Collins, CO.

Bartlein, P.J., K.H. Anderson, P.M. Anderson et al. 1998. Paleoclimate simulations for North America over the last 21,000 years: features of the simulated climate and comparisons with paleoenvironmental data. Quaternary Science Reviews 17:549-585.

Bartos, D.L. 2001. Landscape dynamics of aspen and conifer forests. *In* Sustaining aspen in western landscapes. Symposium proceedings, RMRS-P-18. Fort Collins CO.

Bayard de Volo, S., R.T. Reynolds, S.A. Sonsthagon, S.L. Talbot & M.F. Antolin. 2013. Phylogeography, Post-glacial gene flow and population history of North American northern goshawks (*Accipiter gentilis*). The Auk 140:342-354.

Beier, P., E.C. Rogan, M.F. Ingraldi & S. Rosenstock. 2007. Does forest structure affect reproduction of northern goshawks in ponderosa pine forests? Journal of Applied Ecology 45:342-350.

Benkman, C.W., R.P. Balda & C.C. Smith. 1984. Adaptations for seed dispersal and the compromises due to seed predation in limber pine. Ecology 65:632-642.

Benson, L. 1948. A treatise on the North American Ranunculi. The American Midland Naturalist 40:1-261.78

Betancourt, J.L. 1984. Late quaternary plant zonation and climate in southeastern Utah. Great Basin Naturalist 44:1-35.

Betancourt, J.L. 1990. Late Quaternary biogeography of the Colorado Plateau. *In* Betancourt, J.L, T.R. Van Devender & P.S. Martin, eds., Packrat Middens, the last 40,000 years of biotic change. The University of Arizona Press, Tucson AZ.

Betancourt, J.L. 2004. Advances in arid lands paleobiogeography: the rodent midden record in the Americas. *In* Lomolino, M.V. & L.R. Heaney, eds., Frontiers of biogeography: new directions in the geography of nature. Cambridge University Press.

Billings, W. D. 1974. Adaptations and origins of alpine plants. Arctic & Alpine Research 6:129-142.

Bills, D.J., M.E. Flynn & S.A. Monroe. 2007. Hydrogeology of the Coconino Plateau and Adjacent Areas, Coconino and Yavapai Counties, Arizona. US Geological Survey Scientific Investigations Report 2005-5222.

Blakey, R. & W. Ranney. 2008. Ancient Landscapes of the Colorado Plateau. Grand Canyon Association, Grand Canyon AZ.

Bonnicksen, T.B. 2000. America's ancient forests: from the Ice Age to the age of discovery. Wiley, New York.

Brady, L.F. 1936. The arroyo of the Rio de Flag: a study of an erosion cycle. Plateau 9:33-37. Museum of Northern Arizona, Flagstaff AZ.

Brennan, T.C. & A.T. Holycross. 2009. Amphibians and reptiles in Arizona. Arizona Game and Fish Department, Phoenix AZ.

Bressler, E.J. 2003. Genetic structure and variation of isolated Douglas-fir populations in southwestern North America and northern Mexico. Master's Thesis, NAU, Flagstaff AZ.

Brown, D.E., (ed.). 1994. Biotic communities: Southwestern United States and northwestern Mexico. University of Utah Press, SLC.

Brown, D.E. 2009. Arizona wildlife: the territorial years, 1863-1912. Arizona Game & Fish Department. Phoenix AZ.

Cantor, L.F. & T.G. Whitham. 1989. Importance of below-ground herbivory: pocket gophers may limit aspen to rock outcrop refugia. Ecology 70:962-970.

Carothers, S.W., J.R. Haldeman & R.P. Balda. 1973. Breeding birds of the San Francisco Mountain area and the White Mountains, Arizona. Northern Arizona Society of Science and Art, Inc. Flagstaff AZ.

Carothers, S.W., J.R. Haldeman, R.P Balda & G. Foster. 1978. Recent changes in the status and distribution of some northern Arizona birds. MNA Report, Flagstaff AZ.

Carrara, P.E. 2011. Deglaciation and postglacial treeline fluctuation in the northern San Juan Mountains, Colorado. USGS Professional Paper 1782.

Chambers, J.C. 1991. Patterns of growth and reproduction in a perennial tundra forb (*Geum rossii*): affects of clone area and neighborhood. Canadian Journal of Botany 69:1977-1983.

Chong, G.W., S.E. Simonson, T.J. Stohlgren & M.A. Kalkhan. 2001. Biodiversity: aspen stands have the lead but will nonnative species take over? *In* W.D. Shepperd, D. Brinkley, D. Bartos, T.H. Stohlgren & L.G. Eskew, eds., Sustaining aspen in western landscapes: symposium Proceedings. USDA USFS RMRS-P-18.

Christie, K. 2006. Vascular flora and floristic analysis of the lower San Francisco Volcanic Field, Coconino County, Arizona. Masters Thesis, NAU. Flagstaff AZ.

Chronic, H. & L. Chronic 2004. Pages of stone: Geology of the Grand Canyon and Plateau Country National Parks and Monuments, 2nd ed. The Mountaineers Books, Seattle WA.

Clark, T.A. 2012. Bryophyte Floristics and Ecology in Grand Canyon National Park. Natural Resource Technical Report NPS/SCPN/NRTR—2012/602.

Cline, P. 1982. They came to the mountain. Northern Arizona University & Old Town Press, Flagstaff AZ.

Cocke, A.E., P.Z. Fule & J.E. Crouse. 2005. Forest change on a steep mountain gradient after extended fire exclusion: San Francisco Peaks,

Arizona, USA. Journal of Applied Ecology 42:814-823.

Colton, H.S. 1958. Precipitation about the San Francisco Peaks, Arizona. Museum of Northern Arizona Technical Series No. 2.

Colton, H.S. 1967. The basaltic cinder cones and lava flows of the San Francisco Mountain volcanic field. Museum of Northern Arizona Bulletin 10.

Conway, F.M., D.A. Ferrill, C.M. Hall, A.P. Morris, J.A. Stamatakos, C.B. Connor, A.N. Halliday & C. Condit. 1997. Timing of basaltic volcanism along the Mesa Butte Fault in the San Francisco Volcanic Field, Arizona, from $^{40}Ar/^{39}Ar$ dates: implications for longevity of cinder cone alignments. J. Geophys. Res. 102:815-824.

Cook, B.I., T.R. Ault & J.E. Smerdon. 2015. Unprecedented 21st century drought risk in the American West and Central Plains. Sci. Adv. 1:1-7.

Coons, J.S. 1984. Elevational zonation and monthly changes in the bird community on San Francisco Mountain, Arizona. Master's Thesis, Northern Arizona University, Flagstaff AZ.

Cosner, O.J. 1962. Ground water in the Wupatki and Sunset Crater National Monuments, Coconino County, Arizona. USGS Water Supply Report W 1475-J, p. 357-374.

Coville, F.V. & D.T. MacDougal. 1903. Desert Botanical Laboratory. Carnegie Institution, publication 6. Washington, D.C.

Cowen, R. 1976. The history of life. Wiley, New York.

Cox, B.C. & P.D. Moore. 2005. Biogeography: and ecological and evolutionary approach. Wiley, New York.

Craig, T.P., P.W. Price, K.M. Clancy, G.L. Waring & C.F. Sacchi. 1988. Forces preventing coevolution in the three-trophic-level system: willow, a gall-forming herbivore, and parasitoid. *In* Spencer, K.C., ed., chemical mediation of coevolution. Academic Press, New York.

Crandell, D. R. 1989. Gigantic debris avalanche of Pleistocene age from Mount Shasta volcano, California and debris avalanche hazard zonation. USGS Publication 1861. Denver CO.

Crisafulli, C.M., F.J. Swanson & V.H. Dale. 2005. Overiew of ecological responses to the eruption of Mount St. Helens: 1980-2005. *In* V.H. Dale, F.J. Swanson & C.M. Crisafulli, eds., Ecological responses to the 1980 eruption of Mount St. Helens. Springer, New York.

Damon, P.E., M. Shafiqullah & J.S. Levanthal. 1974. K-Ar chronology for the San Francisco volcanic field and rate of erosion of the Little Colorado River, *In* Karlstrom, T.N., G.A. Swan and R.L. Eastwood, eds., Geology of northern Arizona, Part 1—regional studies. Geological Society of America, Rocky Mountain Section Meeting. Flagstaff AZ. Guidebook, p. 221-235.

Daubenmire, R.F. 1938. Merriam's life zones in North America. Quarterly Review of Biology 13:327-332.

Davis, O.K. & T.E. Moutoux. 1998. Tertiary and Quaternary vegetation history of the Great Salt Lake, Utah, USA. Journal of Paleolimnology 19:417-427.

De Groot, P. & J.J. Turgeon. 1998. Insect-pine interactions. *In* D.M. Richardson, ed., Ecology and biogeography of *Pinus*. Cambridge University Press.

DeByle, N.V. 1985. Wildlife. *In* N.V. DeByle & R.P. Winokur, eds. Aspen: Ecology and Management in the Western United States. USDA USFS RMRS GTR-RM-119.

Dennis, M.D. 1981. Stratigraphy and petrography of pumice deposits near Sugarloaf Mountain, northern Arizona: Flagstaff AZ, Northern Arizona University, Masters Thesis.

DePinte, D.E. 2016. Cone and Seed Insects of Southwestern White Pine, *Pinus strobiformis*, in Arizona, New Mexico, and Texas. Master's Thesis. Northern Arizona University, Flagstaff AZ.

DesRochers, A. & V.J. Lieffers. 2011. Root biomass of regenerating aspen (*Populus tremuloides*) stands of different densities in Alberta. Canadian Journal of Forestry Research 31:1012-1018.

Dexter, L.R. 1981. Snow avalanches on the San Francisco Peaks. Master's Thesis, Northern Arizona University, Flagstaff AZ.

Dickson, B.G., T.D. Sisk, S.E. Sesnie, R.T. Reynolds, S.S. Rosenstock, C.D. Vojta, M.F. Ingraldi & J.M. Rundell. 2014. Integrating single-species management and landscape conservation using regional habitat occurrence models: the northern goshawk in the Southwest, USA. Landscape Ecology 29:803-815.

Dieterich, J.H. 1980. Chimney Spring forest fire history. USDA. For. Serv. Rocky Mountain For. Range Exp. Station Research Paper RM-220.

Dieterich, J.H. 1983. Fire history of southwestern mixed conifer: a case study. Forest Ecology and Management 6:13-31.

Duffield, W.A. 2003. Volcanoes of Northern Arizona. Grand Canyon Association, Grand Canyon AZ.

Eggler, W.A. 1966. Plant succession on the recent volcano, Sunset Crater. Plateau 38:81-96. Museum of Northern Arizona, Flagstaff AZ.

Elias, S.A. 2001. Paleoecology and late Quaternary environments of the Colorado Rockies. *In* Bowman, W.D. & T.R. Seastedt, eds., Structure and Function of an Alpine Ecosystem. Oxford University Press, New York.

Elliott, D. & R.C. Blakey. 2005. The Paleozoic Vertebrates of Arizona. *In* R.D. McCord, ed., Vertebrate Paleontology of Arizona. Mesa Southwest Museum Bulletin No. 11.

Etter, M.P. 2006. Snow sublimation and energy balance along a forest-edge transect in northern Arizona. MA Thesis, Northern Arizona University, Flagstaff AZ.

Fairweather, M.L. 1999. Aspen defoliation on the San Francisco Peaks and Kendrick Mountain. Site Visit Report submitted to Coconino and Kaibab National Forests, June 1999. USDA USFS Coconino National Forest, Flagstaff AZ.

Fairweather, M.L. 2012. Post-fire regeneration and persistence, 2010 Schultz Fire area, Flagstaff Ranger District, Coconino National Forest. Forest Health Report. USFS Southwestern Region. AZ-FHP-12-02.

Fairweather, M.L. & B.W. Geils. 2011. First report of the white blister rust pathogen, *Cronartium ribicola*, in Arizona. Plant Disease 95:494-495.

Fairweather, M.L., B.W. Geils & M. Manthei. 2007. Aspen decline on the Coconino National Forest. Western International Forest Disease Work Conference 55:53-62.

Fawcett, P.J., J.F. Werne, R.S. Anderson, J.M. Heikoop, E.T. Brown, M.A. Berke, S.J. Smith, F. Goff, L. Donohoo-Hurley, L.M. Cisneros-Dozal, S. Schouten, J.S. Sinninghe Damste, Y. Huang, J. Toney, J. Fessenden, G. Woldegbriel, V. Atudorei, J.W. Giessman & C.D. Allen. 2011. Extended megadroughts in the southwestern United States during Pleistocene interglacials. Nature 470:518-521.

Feth, J.H. 1953. A geological and geophysical reconnaissance of the Doney Park-Black Bill Park area, with reference to ground water. USGS Circular 233.

Fisher, R.F. & D. Binkley. 2000. Ecology and management of forest soils. Wiley and Sons, New York.

Fisher, M.A. & P.Z. Fulé. 2004. Changes in forest vegetation and arbuscular mycorrhizae along a steep elevation gradient in Arizona. Forest Ecology and Management 200: 293-311.

Fowler, J.F. & C.H. Sieg. 2010. Density and elevational distribution of the San Francisco Peaks ragwort, *Packera franciscana* (Asteraceae), a single mountain endemic. Madroño 57:213-219.

Froyland, H. 2012. Snow loss on the San Francisco Peaks: effects of an elevation gradient on evapo-sublimation. Master's Thesis, Northern Arizona University, Flagstaff AZ.

Fulé, P.Z., W.W. Covington, H.B. Smith, J.D. Springer, T.A. Heinlein, K.D. Huisinga & M.M. Moore. 2002. Comparing ecological restoration alternatives: Grand Canyon, Arizona. Forest Ecology and Management 170:19-41.

Fulé, P.Z., J.E. Crouse, T.A. Heinlein, M.M. Moore, W.W. Covington & G. Verkamp. 2003*a*. mixed-severity fire regime in a high-elevation forest: Grand Canyon, Arizona. Landscape Ecology 18:465-486.

Fulé, P.Z., T.A. Heinlein, W.W. Covington & M.M. Moore. 2003*b*. assessing fire regimes on Grand Canyon landscapes with fire scar and fire record data. International Journal of Wildland Fire 12:129-145.

Furniss, R.L. & V.M. Carolin. 1977. Western forest insects. Misc. Publ. No. 1339. USFS.

Ganey, J.L. & R.P Balda. 1988. Distribution and habitat use of Mexican spotted owls in Arizona. Condor 91:355-361.

Gardner, R.S. 2013. Clonal diversity of quaking aspen (*Populus tremuloides*): how multiple clones may add to the resilience and persistence of this forest type. Master's Thesis. Utah State University, Utah.

Garnet, G.N., C.L. Chambers & R.L. Mathiasen. 2004. Use of witches' brooms by Abert squirrels in ponderosa pine forests. Wildlife Society Bulletin 34:467-472.

Geils, B.W., K.E. Humer & R.S. Hunt. 2010. White pine, *Ribes* and blister rust: a review and synthesis. Forest Pathology 40:147-185.

Goodrich, B.A, K.M. Waring & T.E. Kolb. 2016. Genetic variation in *Pinus strobiformis* growth and drought tolerance from southwestern US populations. Tree Physiology 36:1219-1235.

Griffis, K. 1999. Bird use of small aspen stands in a northern Arizona ponderosa pine forest. Master's Thesis, NAU, Flagstaff AZ.

Grissino-Mayer, H.D. & T.W. Swetnam. 2000. Century-scale climate forcing of fire regimes in the American Southwest. The Holocene 10:213-220.

Guerra-de la Cruz, V. 2001. Stand structure and dynamics of isolated *Pseudotsuga* forests in southern North America. Master's Thesis, NAU, Flagstaff AZ.

Gulliford, A. 2000. Sacred objects and sacred places. University of Colorado Press, Boulder CO.

Haldeman, J.R., R.P. Balda & S. Carothers. 1973. Breeding birds of a ponderosa pine forest and a fir, pine, aspen forest in the San Francisco Mountain area, Arizona. Museum of Northern Arizona Technical Series No. 12.

Hall, P.A. 2001. Size and composition of northern Goshawk (*Accipiter gentilis*) home ranges in north-central Arizona. Ph.D. Thesis, Northern Arizona University, Flagstaff AZ.

Hamrick, J.L. 1976. Variation and selection in western montane species. II. Variation within and between populations of white fir on an elevational transect. Theor. Appl. Genet. 47:27-34.

Harsbargen, J. 1994. A Holocene paleoclimatic and environmental record from Stoneman Lake, Arizona. Quaternary Research 42:188-196.

Harper, K.T., D.C. Freeman, W.K. Oster & L.G. Klikoff. 1978. The flora of Great Basin mountain ranges: diversity, sources, and dispersal ecology. *In* Harper, K.T. & J.L. Reveal, eds. Intermountain Biogeography: a symposium. Great Basin Naturalist Memoirs 2:79-103.

Harrington, T.C. & M.J. Wingfield. 1998. Diseases and the ecology of indigenous and exotic pines. *In* D.M. Richardson, ed., Ecology and biogeography of *Pinus*. Cambridge University Press.

Hawksworth, F.G. & D. Wiens. 1996. Dwarf mistletoes: biology, pathology and systematics. Agriculture Handbook. USDA USFS. Washington, D.C.

Hedwall, S.J. 2000. Bird and mammal use of dwarf mistletoe-induced witches' brooms in Douglas-fir in the Southwest. Master's Thesis, NAU, Flagstaff AZ.

Heinlein, T. A., M.M. Moore, P.Z. Fulé & W.W. Covington. 2005. Fire history and stand structure of two ponderosa pine-mixed conifer sites: San Francisco Peaks, Arizona, USA. International Journal of Wildland Fire 14:307-320.

Hereford, R. 1984. Climate and ephemeral stream processes: Twentieth century geomorphology and alluvial stratigraphy of the Little Colorado River, Arizona. Geological Society of America Bulletin 95:654-668.

Hereford, R. 2002. Valley-fill alluvium during the Little Ice Age (ca A.D. 1400-1880), Paria River basin and southern Colorado Plateau, United States. Geological Society of America Bulletin 114:1550-1563.

Hereford, R., R.H. Webb & C.I. Longpré. 2006. Precipitation history and ecosystem response to multidecadal precipitation variability in the Mohave Desert region, 1893-2001. Journal of Arid Environments 67:13-34.

Hereford, R. 2014. Climate variation at Flagstaff, Arizona's Mountain Town: compiled from National Weather Service Data. Unpublished.

Hereford, R., G.E. Bentley & H.C. Fairley. 2014. Precipitation variability of the Grand Canyon Region, 1893 through 2009, and its implications for studying the effects of gullying of Holocene terraces and associated archeological sites in Grand Canyon, Arizona: U.S. Geological Survey Open-File Report 2014-1006.

Hermann, R.K & D.P. Lavender. 1990. *Pseudotsuga menziesii* (Mirb.) Franco. *In* Burns, R.M. & B.H. Honkala, eds., Agriculture Handbook, USDA, Washington, D.C.

Hevly, R. 1964. Pollen analysis of Quaternary archaeological and lacustrine sediments from the Colorado Plateau. Doctoral Dissertation , University of Arizona, Tucson AZ.

Hevly, R. 1985. A 50,000 year record of Quaternary environments: Walker Lake, Coconino County, Arizona. Late Quaternary vegetation and climates of the American Southwest. American Association of Stratigraphic Palynologists Contribution Series 16:141-154.

Hevly, R. & T.N.V Karlstrom. 1974. Southwest paleoclimate and continental correlations. *In* Karlstrom. T.N., G.A. Swann & R.L. Eastwood, eds. Geology of Northern Arizona, with notes on archaeology and paleoclimate. Part 1. Regional studies; guidebook for the GSA RM section meeting. 257-295.

Higgins, D.P. 1998. Leakage simulation from a perched mountain aquifer in the Inner Basin, San Francisco Mountain, Arizona. MS Thesis, Northern Arizona University, Flagstaff AZ.

Hirst, S. 2014. Peaks in Flagstaff had Native names before the San Francisco Mountain. Ask a Ranger:3 July 2014. Arizona Daily Sun. Flagstaff AZ.

Hodnett, J-P., D.K. Elliott & T.J Olson. 2013. A New Basal Hybodont (Chondrichthyes, Hybodontiformes) from the Middle Permian (Roadian) Kaibab Formation, of Northern Arizona. *In* Lucas, S.G., et al. eds., 2013, The Carboniferous-Permian Transition. New Mexico Museum of Natural History and Science, Bulletin 60:103-108.

Hoffmeister, D.F. 1986. Mammals of Arizona. University of Arizona Press, Tucson.

Holm, R.F. 1986. Field guide to the geology of the central San Francisco Volcanic Field, northern Arizona. *In* J.D. Nations, C.M. Conway & G.A. Swann, eds., Geology of central and northern Arizona. Geological Society of America Rocky Mountain Section Guidebook. Flagstaff AZ.

Holm, R.F. 1987. San Francisco Mountain: a late Cenozoic composite volcano in northern Arizona. *In* Beus, S.S., ed., Centennial Field Guide. Geol. Soc. Am. Rocky Mt. Sect. 2:389-392.

Holm, R.F. 1988. Geologic map of San Francisco Mountain, Elden Mountain, and Dry Lake Hills, Coconino County, Arizona. U.S. Geol. Surv. Misc. Invest. Ser. Map I-1663, scale 1:24,000.

Holm, R.F. 2004. Landslide preconditions and collapse of the San Francisco Mountain composite volcano, Arizona, into cold debris avalanches in late Pleistocene. Journal of Geology 112:335-348.

Holm, R.F. 2013. Ask a Ranger: Peaks not a blowout. Arizona Daily Sun, Flagstaff AZ. 27 June, 2013.

Johnson, E.P. & T.J. Ayers. 2013. Hybridization of the San Francisco Peaks rare endemic, *Packera franciscana*, with a lower elevation congener: evidence from morphometric and molecular data. The Plant Press 36:13-15.

Jones, R.A. 1993. The relationship of the annual snowpack to the water yield from the Inner Basin of the San Francisco Peaks, Arizona. SCS-USDA report.

Joyal, T. 2004. Fluvial geomorphic response to Holocene climate change on the Mogollon Rim, Arizona. Master's Thesis, Northern Arizona University, Flagstaff AZ.

Kane, J.M., T.E. Kolb & J.D. McMillin. 2014. Stand-scale tree mortality factors differ by site and species following drought in southwestern mixed conifer forests. Forest Ecology and Management 330:171-182.

Karátson, D., T. Telbisz & B.S. Singer. 2010. Late stage volcano geomorphic evolution of the Pleistocene San Francisco Mountain, Arizona (USA), based on high-resolution DEM analysis and ^{40}AR/ ^{39}AR chronology. Bulletin of Volcanology 72:833-846.

Kay, C.E. 2007. Are lightning fires unnatural? A comparison of aboriginal and lightning ignition rates in the United States. *In* Masters, R.E. & K.E., Proceedings of the 23rd Tall Timbers Fire Ecology Conference: fire in grassland and shrubland ecosystems. Tall Timbers Research Station, Tallahassee, FL.

Kearny, T.H. & R.H. Peebles. 1964. Arizona Flora. University of California Press, Berkeley.

Kelso, S., P.M. Beardsley & K. Weitemier. 2009. Phylogeny and biogeography of Primula sect. Parryi (Primulaceae). International Journal of Plant Sciences. 170:93-106.

Kerhoulas, L.P., T.E. Kolb, M.D. Hurteau & G.W. Koch. 2013*a*. Managing climate change adaptation in forests: a case study from the U.S. Southwest. Journal of Applied Ecology 50:1311-1320.

Kerhoulas, L.P., T.E. Kolb & G.W. Koch. 2013*b*. Tree size, stand density, and the source of water used across seasons by ponderosa pine in northern Arizona. Forest Ecology and Management 289:425-433.

Kruckeberg, A.R. 1987. Plant life on Mount St. Helens before 1980. *In* D.E. Bilderback, ed., Mount St. Helens 1980: botanical consequences of the explosive eruptions. University of California Press, Berkeley CA.

La Marche, V.C. 1973. Holocene climatic variations inferred from treeline fluctuations in the White Mountains, California. Quaternary Research 3:632-660.

Lamb, T., T.R. Jones & P.J. Wettstein. 1997. Evolutionary genetics and phylogeography of tassel-eared squirrels (*Sciurus aberti*). Journal of Mammalogy 87:117-133.

Lanner, RM. 1996. Made for each other: A symbiosis of Birds and Pines. Oxford University Press, New York.

Lanner, R.M. & S.B. Vander Wall. 1980. Dispersal of limber pine seeds by Clark's nutcracker. Journal of Forestry 78:637-639.

Latta, R.G. & J.B. Mitton.1999. Historical separation and present gene flow through a zone of secondary contact in ponderosa pine. Evolution 53:769-776.

Ledbetter, J.D., L.E. Stevens & B. Brandt. 2018. Springs and Springs-dependent Species Online Database. Version 1.0. Springs Stewardship Institute, Museum of Northern Arizona, Flagstaff AZ. Springsdata.org.

Ledig, F.T., P.D. Hodgkiss & D.R. Johnson. 2006. The structure of genetic diversity in Engelmann spruce and a comparison with blue spruce. Can. J. Bot. 84:1806-1828.

Ledig, F.T., P.D. Hodgkiss, K.V. Krutovskii, D.B. Neale, & T. Eguiluz-Piedra. 2004. Relationships among the Spruces (*Picea*, Pinaceae) of southwestern North America. Systematic Botany 29:275-295.

Leiberg, J.B., T.F. Rixon & A. Dodwell. 1904. Forest conditions of the San Francisco Mountains Forest Reserve, Arizona. USGS Professional Paper No. 32, Series H, Forestry 7. Washington, D.C.

Li, P. & W.T. Adams. 1989. Range-wide patterns of allozyme variation in Douglas-fir (*Psuedotsuga menziesii*). Canadian Journal of Forest Research 19:149-161.

Lindsay, E. 1978. Late Cenozoic vertebrate faunas, southeastern Arizona. New Mexico Geological Society Guidebook. 29th Field Conference, Land of Cochise, New Mexico.

Lisiecki, L.E. & M.E. Roymo. 2005. A Pliocene-Pleistocene stack of 57 globally-distributed benthic d^{18}O records. Paleoceanography 20, PA1003, doi:10.1029/2004PA001071.

Little, E. L. 1941. Alpine flora of San Francisco Mountain, Arizona. Madroño 6:65-96.

Lomolino, M.V., B.R. Riddle, R.J. Whittaker & J.H. Brown. 2006. Biogeography. Sinaurer and Associates, MA.

Looney, C.E. & K.M. Waring. 2012. Patterns of forest structure, competition and regeneration in southwestern white pine (*Pinus strobiformis*) forests. Forest Ecology & Management 286: 159-170.

Lynch, A.M., J.A. Anhold, J.D. McMillin, S.M. Dudley, R.A. Fitzgibbon & M.L. Fairweather. 2008. Forest insect and disease activity on the Coconino National Forest, 1918-2006.

MacDougal, W.B. 1973. Seed plants of northern Arizona. Museum of Northern Arizona, Flagstaff AZ.

Macdougall, D. 2013. Frozen Earth, the once and future story of ice ages. University of California Press, Berkeley CA.

Margolis, E.Q., T.W. Swetnam & C.D. Allen. 2011. Historical stand-replacing fire in upper montaine forests of the Madrean Sky Islands and Mogollon Plateau, Southwestern USA. Fire Ecology 7:88-107.

Martin, E.C. 1969. The relation of precipitation and flow from Little Leroux Spring on the SF Peaks, Arizona. USDA-FS Research Note RM-129.

Masek, Lopez, S. & A. Springer. 2002. Assessment of human influence on riparian change in the Verde Valley, Arizona. Fulfillment of Water Protection Fund Grant 98-057.

Mast, J.N., P.Z. Fulé, M.M. Moore, W.W. Covington & A. Waltz. 1999. Restoration of presettlement age structure of an Arizona ponderosa pine forest. Ecological Applications 9:228-239.

Mattson, D.J. 2007. Mountain lions of the Flagstaff uplands, 2003-2006. Open file report 2007-1062. USGS.

Mattson, D.J. & T.A. Arundel. 2013. Consumption of seeds of southwestern white pine (*Pinus strobiformis*) by black bear (*Ursus americanus*). Southwestern Naturalist 58:243-245.

McCabe, L.M. 2015. The transition of bee to fly pollinator communities along an elevational gradient. Masters Thesis, NAU, Flagstaff AZ.

McCabe, L.M., P. Cheshire, D. Smith & N.S. Cobb. 2018. Bees of the San Francisco Peaks, AZ. SCAN checklist.

McCleary, J.A. 1962. The distributional studies of Arizona mosses. The American Midland Naturalist 67:68-78.

McKee, E.H., P.E. Damon, M. Shafiqullah, R.C. Harris & J.E. Spencer. 1998. Compilation of unpublished USGS and University of Arizona K-Ar dates of volcanic rocks of the San Francisco volcanic field, northern Arizona. Ariz. Geol. Surv. Open File 98-2.

McKelvey, S.R. 2004. Comparison of two methods of site index curve construction for white fir in the Southwest. Master's Thesis, NAU, Flagstaff AZ.

McMillin, J.D. 2005. Douglas-fir beetle activity on the San Francisco Peaks. USDA-USFS Site Visit Report. Submitted to Peaks RD, Coconino National Forest, Arizona.

McMillin, J.D. 2006. Spruce beetle activity at Snowbowl Ski Area. USDA-USFS Site Visit Report. Submitted to Peaks RD, Coconino National Forest, Arizona.

Mead, J.I. & L.D. Agenbroad. 1992. Isotope dating of Pleistocene dung deposits from the Colorado Plateau, Arizona and Utah. Radiocarbon 34:1-19.

Menon, M., J.C. Bagley, C. Friedline, A.V. Whipple, A.W. Schoettle, A. Leal-Seanz, C. Wehenkel, F. Molina-Freaner, L. Flores-Renteria, M. Socorro Gonzales-Elizondo, R.A. Sneizko, S.A. Cushman, K.M. Waring & A.J. Eckert. 2018. The role of hybridization during ecological divergence of southwestern white pine (*Pinus strobiformis*) and limber pine (*Pinus flexilis*). Doi:https/doi.org/ 10.1101/185728.

Merriam. C.H. 1890. Results of a biological survey of the San Francisco Mountain region and desert of the Little Colorado in Arizona. USDA Div. Ornith. & Mammal. North American Fauna 3. 136 pp.

Metcalfe, J.Z., F.J. Longstaffe & G. Hodgins. 2013. Proboscideans and paleoenvironments of the Pleistocene Great Lakes: landscape, vegetation, and stable isotopes. Quaternary Science Reviews 76:102-113.

Michalet, R., Xaio, S., Touzard, B., Smith, D.S., Cavieres, L.A., Callaway, R.M. & T. Whitham. 2011. Phenotypic variation in nurse traits and community feedbacks define an alpine community. Ecology Letters 14:433-443.

Millar, C.I. 1998. Early evolution of pines. *In* D.M. Richardson, ed., Origin and evolution of Gymnosperms. Cambridge University Press, Cambridge.

Miller, J., M. Litvak, S. Kelso & A. Vargo. 1994. Comparative reproductive biology of two alpine primrose species. Arctic and Alpine Research 26:297-303.

Mitton, J.B. & M.C. Grant. 1996. Genetic variation and the natural history of quaking aspen. Bioscience 46:25-31.

Mitton, J.B., B.R. Kreiser & R.G. Latta. 2000. Glacial refugia of limber pine (*Pinus flexilis* James) inferred from the population structure of mitochondrial DNA. Molecular Ecology 9:91-97.

Moir, W.H. 1993. Alpine tundra and coniferous forest. *In* W.A. Dick-Peddie, ed., New Mexico vegetation. University of New Mexico Press, New Mexico.

Moir, W.H. & J.A. Ludwig. 1979. A classification of spruce-fir and mixed conifer habitat types of Arizona and New Mexico. USDA USFS Research Paper RM-207.

Moir, W.H., B. Geils, M.A. Benoit & D. Scurlock. 1997. Ecology of southwestern ponderosa pine forests. USDA USFS RM-GTR-292.

Moir, W.H. & Deaver Herbarium Staff NAU. 2005. Vascular plants of the San Francisco Peaks Area. Deaver Herbarium, Northern Arizona University, Flagstaff AZ. Online version available.

Monroe, S., J. Grahame & S. Clark. 2013. Leroux Redux—release of flow at Big Leroux Spring near Flagstaff, Arizona. 12th Biennial Conference of Science and Management on the Colorado Plateau. NAU, Flagstaff AZ.

Mooney, K.A. 2006. The distruption of an ant-aphid mutualism increases the effects of insectivorous birds on pine herbivores. Ecology 87:1805-1815.

Moore, R.B. 1974. Geology, Petrology and Geochemistry of the eastern San Francisco Volcanic Field, Arizona: PhD. Thesis. University of New Mexico, Albuquerque, NM.

Moore, T.C. 1965, origin and disjunction of the alpine tundra on the San Francisco Mountain, Arizona. ecology 46:860-864.

Morgan, P., J.H. Sass & W. Duffield. 2010. Geothermal resource evaluation program of the eastern San Francisco Volcanic Field. Arizona Geological Survey Contributed Report CD-10-D, Phoenix AZ. www.azgs.gov.

Morris, W. & D. Doak. 1998. Life history of the long-lived gynodioecious cushion plant *Silene acaulis* (Caroyphyllaceae), inferred from size-based population projection matrices. American Journal of Botany 85:784-793.

Moser, L. P. 1999. Genetic structure and variation in a southwestern Douglas-fir population. Master's Thesis, NAU, Flagstaff AZ.

Nash, T.H. 2002. Lichen flora of the greater Sonoran Desert region. Arizona State University Press, Tempe AZ.

Neff, L.T., K.C. Anderson, D.A. Anderson, K. Spur, T. Joyal & R. Richardson. 2003. Alluvial stratigraphy, landscape evolution, and archaic period occupation along Walnut Canyon drainage and adjacent areas: new research regarding landscape change, climatic inferences and human occupation. 7th Biennial conference Integrating Science and Management on the Colorado Plateau, November 2003, Flagstaff AZ.

Nowicki, B. 2003. Genetic diversity and dwarf mistletoe resistance in a southwestern Douglas-fir population. Master's Thesis, NAU, Flagstaff AZ.

Olberding, S.D. 2002. Fort Valley then and now: a look at an Arizona settlement. Fort Valley Publishing, Flagstaff AZ.

Paklaian, J.R. 2017. Holocene and late Pleistocene paleoecological history using lake sediment in the Chuska Mountains, Navajo Nation, New Mexico. Master's Thesis, Northern Arizona University, Flagstaff AZ.

Paulik, L.A. 1980. A Vascular Flora of the Subalpine Spruce-Fir Forest of the San Francisco Peaks. Masters Thesis, Northern Arizona University, Flagstaff AZ.

Peacock, D. 2013. In the shadow of the sabertooth: global warming, the origin of the first Americans and the terrible beasts of the Pleistocene. AK Press. Oakland CA.

Pearson, G.A. 1920. Factors controlling the distribution of forest types, Part I. Ecology 1:139-159.

Pearson, G.A. 1920. Factors controlling the distribution of forest types, Part II. Ecology 1: 289-308.

Pearson, G.A. 1923. Natural reproduction of western yellow pine in the Southwest. USDA Bulletin No. 1105

Pearson, G.A. 1931. Forest types in the Southwest as determined by climate and soil. USDA Technical Bulletin No. 247.

Pearson, L.C. & D.B. Lawrence. 1958. Photosynthesis in aspen bark. American Journal of Botany 45:383-387.

Peters, L. 2003. ^{40}AR/^{39}AR Geochronology Results: New Mexico Geochronological Laboratory Research Internal Report #NMGRL-IR-360. Also available as an appendix in Morgan et al. 2010.

Péwé, T.L. & R.G. Updike. 1976. San Francisco Peaks: a guidebook to the geology. Second Edition. Museum of Northern Arizona, Flagstaff AZ.

Phillips, A.R. 1939. The type of *Empidonax wrightii* Baird. Auk 56:311-312.

Pilles, P.J. 1979. Sunset Crater and the Sinagua: A new interpretation. *In* Sheets, P.D. & D.K. Grayson, eds., Volcanic activity and human ecology. Academic Press, New York.

Poellot, C.D. 2000. Ice surface reconstruction and energy balance modelling of alpine glaciers. Master's Thesis, Oregon State University, Oregon.

Priest, S.S., W. A. Duffield, N. R. Riggs, B. Porturalski and K. Malis-Clark. 2002. Red Mountain Volcano, a spectacular and unusual cinder cone in northern Arizona. USGS Fact Sheet 024-02.

Pyne, S.J. 2001. Fire: a brief history. University of Washington Press, Seattle.

Railsbacks, L.B, P.L. Gibbard, M.J. Head, N.R.G. Voarintsoa, & S. Toucanne. 2015. An optimized scheme of lettered marine isotope substages for the last million years, and the climatostratigraphic nature of isotopic stages and substages. Quaternary Science Reviews 111:94-106.

Rehfeldt, G.E. 1990. Genetic differentiation among populations of *Pinus ponderosa* from the upper Colorado River Basin. Botanical Gazette 151:125-137.

Rehfeldt, G.E. 1993. Genetic variation in the Ponderosae of the Southwest. American Journal of Botany 80:330-343.

Rehfeldt, G.E. 2004. Interspecific and intraspecific variation in *Picea engelmannii* and its congeneric cohorts: biosystematics, genecology, and climate change. General technical Report RMRS-GTR-134. USFS.

Rehfeldt, G.E., D.E. Ferguson& N.L. Crookston. 2009. Aspen, climate and sudden decline in western USA. Forest Ecology and Management 258:2353-2364.

Rehfeldt, G.E., B.C. Jaquish, J. Lopez-Upton, C. Sáenz-Romero, J.B. St. Clair, L.P. Leites & D.G. Joyce. 2014. Comparative genetic responses to climate for the varieties of *Pinus ponderosa* and *Pseudotsuga menziesii*: realized climate niches. Forest Ecology and Management 324:126-137.

Reveal, J.L. 1979. The Intermountain region: its biogeography and its evolutionary history. Mentzelia 4:1-92.

Riggs, N.R., T.M. Lehman, G.E. Gehrels & W.R. Dickinson. 1996. Detrital zircon link between headwaters and terminus of the Upper Triassic Chinle-Dockum paleoriver system. Science 273:97-100.

Ritchie, J.C. & G.M. MacDonald. 1986. The patterns of post-glacial spread of white spruce. Journal of Biogeography 13:527-540.

Robinson, H.H. 1913. The San Francisco Volcanic Field, Arizona. U.S. Geological Survey Professional Paper 76.

Romanowski, S. & S.T. Jackson, eds. 2009. Alexander von Humboldt's essay on the geography of plants. University of Chicago Press. Chicago IL.

Romme, W.H., M.G. Turner, L.L. Wallace & J.S. Walker. 1995. Aspen, elk and fire in Northern Yellowstone Park. Ecology 76:2097-2106.

Rosenstock, S.S. 1998. Influence of Gambel oak on breeding birds in ponderosa pine forests of northern Arizona. Condor 100:485-492.

Salmon, O. & D. McAvoy. 2009. Pando sheds light on genetic diversity in aspen stands. Utah Forest News 13:1, Winter. Utah State Extension Service.

Salzer, M.W. & K.T. Kipfmueller. 2005. Reconstructed temperature and precipitation on a millennial timescale from tree-rings in the southern Colorado Plateau, U.S.A. Climate Change 70:465-487.

Schaack, C.G. 1970. A flora of the arctic-alpine vascular plants of the San Francisco Mountain. Masters Thesis, Northern Arizona University, Flagstaff AZ.

Schaack, C. G. 1983. The alpine vascular flora of Arizona. Madroño 30:79-88.

Scott, J.M. 2008. Western spruce dwarf mistletoe: taxonomic investigations and the effects of parasitism on bristlecone pine. Masters Thesis, Northern Arizona University, Flagstaff AZ.

Scott, J.M. & R. L. Mathiasen. 2009. Bristlecone pine dwarf mistletoe: *Arceuthobium microcarpum* subsp. *aristatae* (Viscacaeae): a new species of western spruce dwarf mistletoe from northern Arizona. Journal of the Botanical Research Institute of Texas. 3:13-21.

Searcy, B.T., S.M. Beckstrom-Sternberg, J.S. Beckstrom-Sternberg, P. Stafford, A.L. Schwendiman, J. Soto-Pena, M.C. Owen, C. Ramirez, J. Phillips, N. Veldhoen, C.C. Helbing & C.R. Propper. 2012. Thyroid hormone-dependent development in *Xenopus laevis*: a sensitive screen of thyroid hormone signaling disruption by municipal wastewater treatment plant effluent. General and Comparative Endochronology 176:481-492.

Seager, R., M. Ting, I. Held, Y. Kushnir, J. Lu, G. Vecchi, H-P. Huang, N. Harnik, A. Leetmaa, N-C. Lau, C. Li, J. Velez & N. Naik. 2007. Model projections of am imminent transition to a more arid climate in southwestern North America. Science 316:1181-1184.

Sellers, W.D. & R.H. Hill. 1985. Arizona climate, 1885-1985. Institute of Atmospheric Sciences, University of Arizona, Tucson AZ.

Sharp, R.P. 1942. The Pleistocene glaciation of San Francisco Mountain, Arizona. New Mexico Geological Society 9th Annual Fall Field Conference Guidebook, pages 151-152.

Sheridan, M.F & R.G. Updike. 1975. Sugarloaf Mountain tephra—a Pleistocene rhyolitic deposit of base-surge origin in northern

Arizona. Geological Society of American Bulletin 86:571-581.

Shirk, A.J., S.A. Cushman, K.M. Waring & C.A. Wehenkel. 2018. Southwestern white pine (*Pinus strobiformis*) species distribution models project a large range shift and contraction due to regional climate change. Forest Ecology and Management 411:176-186.

Sibley, D.A. 2003. The Sibley field guide to birds of western North America. Knopf, New York.

Spence, J.R., T.A. Clark & J. Brinda. [In Preparation] Checklist of the bryophytes of Arizona.

Spence, J.R., W.H. Romme, L. Floyd-Hanna & P.G. Rowlands. 1995. A preliminary vegetation classification for the Colorado Plateau. *In* C. Van Riper, ed., Proceedings on the second biennial conference on research in Colorado Plateau National Parks. NPS Trans. Proc. NPS/NRNAU/NRTP-95/11.

Spence, J.R. 2011. The Spence/Romme/Floyd-Hanna/Rowlands classification (SRFR). Version 5.0. NPS.

Springer, J.D., M.L. Daniels & M. Nazaire. 2009. Field guide to forest & mountain plants of Northern Arizona. Northern Arizona Ecological Restoration Institute, NAU, Flagstaff AZ.

States, J.S. 1990. The mushrooms and truffles of the Southwest. Transition Zone Horticultural Institute, Flagstaff AZ.

States, J. S. & W.S. Gaud. 1997. Ecology of hypogeous fungi associated with ponderosa pine. I. patterns of distribution and sporocarp production in some Arizona forests. Mycologia 89:712-721.

Stempniewicz, V.A. 2014. Evaluating erosion risk mitigation due to forest restoration treatments using alluvial chronology and hydrologic modeling. Masters Thesis, NAU, Flagstaff AZ.

Stevens, L.E., T.L. Griswold, O. Messenger, W.G. Abrahamson & T.J. Ayers. 2007. Plant and pollinator diversity in northern Arizona. The Plant Press-Arizona Native Plant Society. April:5-8.

Stevens, L.E. & R.L. Huber. 2004. Biogeography of tiger beetles (Cicindelidae) in the Grand Canyon Ecoregion, Arizona and Utah. Cicindela 35:41-64.

Stevens, L.E., F.B. Ramberg & R.F. Darsie, Jr. 2008. Biogeography of Culicidae (Diptera) in the Grand Canyon region, Arizona, USA. Pan-Pacific Entomologist 84:92-109.

Swanson, F.J. & J.J. Major. 2005. Physical events, environments and geological-ecological interactions at Mount St. Helens: March 1980-2004. *In* V.H. Dale, F.J. Swanson & C.M. Crisafulli, eds., Ecological responses to the 1980 eruption of Mount St. Helens. Springer Science+Business Media, Inc.

Swetnam, T.W. & C.H. Baisan. 2003. Tree-ring reconstructions of fire and climate history in the Sierra Nevada and southwestern United States. *In* Veblen, T.T., et al., Fire and climatic change in temperate ecosystems of the western Americas. Springer, New York.

Taylor, W.P. 1935. Ecology and life history of the porcupine (Erithizon epizanthum) as related to the forests of Arizona and the southwestern United States. University of Arizona Press, Tucson AZ.

Tomback, D.F. 1988. Nutcracker-pine mutualism: Multitrunk trees and seed size. Acta International Ornithological Congress XIX:518-527.

Ulrich, G.E., G.H. Billingsly, R. Hereford, E.W. Wolfe, L.D. Nealy & R.L. Sutton. 1984. Map showing geology, structure, and uranium deposits of the Flagstaff quadrangle, Arizona: USGS Survey Map I-1446.

Updike, R.G. 1977. The geology of the San Francisco Peaks, Arizona. PhD. Dissertation, Arizona State University, Tempe AZ.

Updike, R.G. & T.L. Péwé. 1970. A new quaternary formation in northern Arizona. Plateau. 43:21-26. Museum of Northern Arizona, Flagstaff AZ.

Van Devender, T.R. 1990. Late Quaternary vegetation and climate of the Sonoran Desert, United States and Mexico. *In* Betancourt, J.L., T.R. Van Devender & P.S. Martin, eds., Packrat Middens: the last 40,000 years of biotic change. University of Arizona Press, Tucson AZ.

Van Devender, T.R. 1995. Desert grassland history: changing climates, evolution, biogeography, and community dynamics. *In* McClaran, M.P. & T.R. Van Devender, eds. The Desert Grassland. University of Arizona Press, Tucson AZ.

Vankat, J.L. 2013. Vegetation dynamics on the

mountains and plateaus of the American Southwest. Springer, New York.

Wallace, A. & R.H. Hevly. 2007. From Texas to to San Diego in 1851: the overland diary of Surgeon Dr. S.W. Woodhouse, surgeon-naturalist of the Sitgreaves expedition. Texas Tech University Press, Lubbock, Texas.

Waring, G.L. 2011. The natural history of the Intermountain West: its ecological and evolutionary story. University of Utah Press, Salt Lake City UT.

Warner, T.T. 2004. Desert meteorology. Cambridge University Press, New York.

Wells, P.V. & J.D. Stewart. 1987. Cordilleran-boreal taiga and fauna on the central Great Plains of North America, 14,000-18,000 years ago. American Midlands Naturalist 118:94-106.

Weng, C & S.T. Jackson. 1999. Late glacial and Holocene vegetation history and paleoclimate of the Kaibab Plateau, Arizona. Paleogeography, Paleoclimatology, Paleoclimatology 153:179-201.

Wenrich-Verbeek, K. 1979. The petrogenesis and trace-element geochemistry of intermediate lavas from Humphreys Peak, San Francisco volcanic field, Arizona. Tectonophysics. 61:103-130.

Whittaker, R.H. & W.A. Nearing 1975. Vegetation of the Santa Catalina Mountains, Arizona. V. Biomass, production and diversity along the elevation gradient. Ecology 56:771-790.

Wolfe, A.D., S.L. Datwyler & C.P. Randle. 2002. A phylogenetic analysis of the Chelonae (Scrophulariaceae) based on ITS and *matK* sequence data. Systematic Botany 27:138-148.

Wolfe, A.D., C.P. Randle, S.L. Datwyler, J.J. Morawetz, N. Arguedas & J. Diaz. 2006. Phylogeny, taxonomic affinities, and biogeography of *Penstemon* (Plantaginaceae) based on ITS and $_{CP}$DNA sequence data. American Journal of Botany 93:1699-1713.

Wolff, S.E., N. Veldhoen, C.C. Helbing, C.A. Ramirez, J.M. Malpas & C.R. Propper. 2015. Estrogenic environmental contaminants alter the mRNA abundance profiles of genes involved in gonadal differentiation of the American bullfrog. Science of the Total Environment 521-522:380-387.

Woodhouse, C.A, K.E. Kunkel, D.R. Easterling & E.R. Cook. 2005. The twentieth-century pluvial in the western United States. Geophysical Research Letters 32:L07701. DOI 1010.129.

Worrall, J.J., G.E. Rehfeldt, A. Hamann, E. H. Hogg, S.B. Marchetti, M. Michaelian & L.K. Gray. 2013. Recent declines of *Populus tremuloides* in North America linked to climate. Forest Ecology & Management 299:35-51.

Zegler, T.J., M.M. Moore, M.L. Fairweather, K.B. Ireland & P.J. Fulé. 2012. *Populus tremuloides* mortality near the southwestern edge of its range. Forest Ecology and Management 282:196-207.

Zhou, B. 1994. Quantifying Mexican spotted owl habitat with the Johnson system of distributions. Master's Thesis, NAU, Flagstaff AZ.

— INDEX —

Page numbers in ***semibold italics*** refer to figures and tables.

Gwendolyn L. Waring is a mother, writer, ecologist and artist based in Flagstaff, AZ. Her writing is driven by a great love of the West, where she has resided for 40 years, after a life of travelling. She explores the region through research and painting. She is the author of another western natural history, *A Natural History of the Intermountain West*, published by University of Utah Press in 2011; it considers the arid corridor that extends through the West, spanning the Colorado Plateau and the Great Basin. Future writing projects include stories of the building of other landscapes in western North America.